News Literacy

Lee B. Becker
GENERAL EDITOR

Vol. 7

The Mass Communication and Journalism series
is part of the Peter Lang Media and Communication list.
Every volume is peer reviewed and meets
the highest quality standards for content and production.

PETER LANG
New York • Washington, D.C./Baltimore • Bern
Frankfurt • Berlin • Brussels • Vienna • Oxford

News Literacy

Global Perspectives for the Newsroom and the Classroom

EDITED BY PAUL MIHAILIDIS

PETER LANG
New York • Washington, D.C./Baltimore • Bern
Frankfurt • Berlin • Brussels • Vienna • Oxford

KH

Library of Congress Cataloging-in-Publication Data

News literacy: global perspectives for the newsroom and the classroom /
edited by Paul Mihailidis.
p. cm. — (Mass communication and journalism; v. 7)
Includes bibliographical references and index.
1. Media literacy. 2. Citizen journalism. 3. Media literacy—Study and teaching.
4. Journalism—Study and teaching—History—21st century. I. Mihailidis, Paul.
P96.M4N5495   302.23—dc23   2011046858
ISBN 978-1-4331-1564-6 (hardcover)
ISBN 978-1-4331-1563-9 (paperback)
ISBN 978-1-4539-0545-6 (e-book)
ISSN 2153-2761

Bibliographic information published by **Die Deutsche Nationalbibliothek.**
Die Deutsche Nationalbibliothek lists this publication in the "Deutsche
Nationalbibliografie"; detailed bibliographic data is available
on the Internet at http://dnb.d-nb.de/.

The paper in this book meets the guidelines for permanence and durability
of the Committee on Production Guidelines for Book Longevity
of the Council of Library Resources.

© 2012 Peter Lang Publishing, Inc., New York
29 Broadway, 18th floor, New York, NY 10006
www.peterlang.com

Printed in the United States of America

7/18/12

This book is dedicated to the Salzburg Academy on
Media & Global Change

TABLE OF CONTENTS

CONTENTS

LIST OF FIGURES, IMAGES, AND TABLES

FIGURES

IMAGES

TABLES

PREFACE

For those interested in global problem-solving, the role of news media cannot be over-estimated. In both democratic societies and in autocratic states, the media's influence today is powerful and not easily controlled. Media provide indispensible means by which citizens form opinions on the causes and consequences of current issues, and on the range of possible solutions. The increasing fragmentation of news audiences makes building consensus more complex but no less essential for concerted public action. Citizens increasingly access and share information instantaneously. As this reality takes hold across the world, the half-life of the "news cycle" shortens, and with it the durability of public opinion. As protest and demonstrations attest, public intolerance grows for governments perceived to be corrupt, unresponsive or inept. This changing relationship among media, public opinion, and political engagement makes urgent the need for reforms and innovations that support effective citizenship.

The forces driving media democratization are also disrupting the economic models of traditional news organizations, limiting their ability to adapt with the speed that circumstances require. The proliferation of hand-held devices – particularly smart phones with photographic, recording, texting and broadband capabilities – means that billions across the planet have the means to produce as well as consume news in real time. New opportunities have opened for digital media entrepreneurs, but it is a formidable task to replace news structures built up over more than a century.

Whatever the contours of the future news paradigm, it is certain to demand a more active role than that played by earlier news consumers. Already, the mixed economy of traditional and social media requires the ordinary citizen to assume greater responsibility for separating fact from fiction, analysis from opinion, and to gain the rudimentary skills necessary to create and edit media. Schools and colleges acknowledge the importance of analytic thinking and developing good communication skills, and the vital role of an educated citizenry in the practice of democracy. But few have done an effective job of preparing their teachers and students to make critical media choices.

In 2007, the Salzburg Global Seminar established a summer Academy to address the changing nature of journalism, and to explore how education could better prepare journalists and ordinary citizens for the digital revolution that was under way. The idea was to create a laboratory for teaching and research, and a network for innovation in journalism and communications that would connect faculty and students from leading universities in every region of the world. Drawing on leading thinkers and practitioners, the Academy takes a comparative approach to: 1) understanding the role that media play in framing global issues and influencing problem-solving, 2) evaluating emerging digital media and news forms in diverse societies, 3) encouraging changes in journalism education that encourage excellence and innovation in digital reporting, and 4) developing a global media literacy curricula to promote more active and discerning news audiences.

The concepts developed in this volume took shape in conversations over four years. The ideas sprang from the minds of the authors, all of whom are leading scholars and teachers in the fields of journalism and communications, and all of whom have participated in at least two sessions of the Salzburg Academy on Media and Global Change. But they were influenced by the interactions among the authors and the Academy's students – remarkable young men and women who helped design, test and refine the notion of what media literacy means, how it can be taught, and how its impact can be evaluated. All of the authors readily acknowledge that they learned at least as much from the students as they may have taught. In that sense, the summer laboratory, the projects that faculty and students took home and the cross border collaborations that ensued, all contributed to this global assessment of *News Literacy*.

After almost 65 years of identifying and convening young leaders from across the world, the Salzburg Global Seminar takes special pride in the work this Academy has produced. More than 250 future editors, researchers, curators and teachers of journalism have gained crucial insight into field they will help shape and lead locally, regionally and internationally; in so doing, they have helped define what an informed citizenry requires from its media in the 21st century, and how schools and universities can equip them to help demand and create the institutions that can deliver on those needs. We trust this book, spawned by active collaboration, will provide a springboard for ongoing thinking in a crucial emerging field of scholarship and teaching. We invite readers to contribute as well as innovate, by joining the Academy community at www.salzburg.umd.edu.

Stephen Salyer, President
Salzburg Global Seminar
www.SalzburgGlobal.org

ACKNOWLEDGMENTS

The idea for this book arose over two years ago at the 2009 Salzburg Academy on Media & Global Change. The faculty group at that session, which comprised most of the authors in this book, wondered over dinner what academic value the Academy could bring in terms of scholarship and contribution to the field beyond its robust curricular and pedagogical scope. The result is this book. This collaborative work reflects the true energy, dynamism, and global reach of the Academy, and would not have been possible without the contribution of a host of dedicated scholars, educators, professionals, and students.

The Academy is a rich web of organizations, institutions, and individuals, who have collectively made it a truly transformative program. Susan Moeller, the Academy's co-founder and chief visionary, has defined the parameters for this book through her leadership and tireless energy in more ways than she knows. I am greatly indebted to her for providing so many opportunities. Jad Melki has worked tirelessly since 2006 to help build the Academy and specifically its research program, without which the Academy would lack a rigorous mode of inquiry. Moses Shumow has vetted various versions of this project at various stages, always providing insightful and detailed criticism, feedback, and direction. The authors of the book, whom you'll meet in the following chapters below, have all brought energy, passion, perspective, and patience, to the Academy over the last five years. If not for their dedication and time, the Academy would not be what it is today. Beyond the authors of this book, faculty have been coming to the Academy for five years, donating their valuable time and energy to the program. They are as integral to this book as those who have penned the chapters herein. A host of other administrators, foundations, faculty and visitors have further offered numerous avenues of support for the Academy. Among them all, special thanks go to: The Knight Foundation, Ford Foundation, UNESCO, Serge Dumont, Karen Tang, Thomas Kunkel, Xiguang Li, Lizette Rabe, Stephen Jukes, Silvia Pellegrini, Ana Pereson, Gabrielle Warkentin, Jorge Liotti, David Burns, Emily Brown, Slavomir Mirgal, Dana Petranova, Denisa Kralovicova, Cecilia Balbin, William Porath, Christian Schwartz, Keith Hughes, Jana Illesova, Roman Gerodimos, Zhu Lian, Zhou Qingan, Noman Sattar, George Papagiannis, Clement So, Liz Lufkin, Patrick Cooper, Tegan Bedsger, Cory Haik, Mohammed Shamma, Sawsan Zaidah, Carol Reese, and Amy Mihailidis.

For over 65 years, the Salzburg Global Seminar has been convening young leaders from around the world to find solutions to the challenges that define the present global landscape. Stephen Salyer, President and CEO of the Seminar, has been instrumental in providing leadership, support, and freedom for the Academy to grow into the dynamic program it is. Jochen Fried, who directs education initiatives for the Seminar, co-directed the Academy program and has remained a steadfast presence in its direction and scope. On the program side, Edward Mortimer, David Goldman, Daniel Sip, Clare Boyle, Andrea Lopez-Portillo, Cheryl Van Emberg, and a host of others have helped to develop the logistics of the Academy and its connection to the larger mission of the Seminar. On the support side, Meg Harris, Patricia Benton, and Lynn McNair, among others, have driven the wide and far-reaching support structure needed to maintain the truly global scope of the program.

I also wanted to thank Mary Savigar, the editor for this book at Peter Lang, for helping to make the project a reality. Also thanks to University of Georgia professor Lee Becker, who accepted this book as part of his journalism and mass community series for Peter Lang. This book builds on the growing presence of the news literacy movement and those who have come to understand its place in the large media literacy landscape. Thanks to all the media literacy scholars worldwide, particularly Renee Hobbs, David Buckingham, and Henry Jenkins, who have helped lay the foundation and frameworks for this book to grow. Specific thanks to Howard Schneider and Dean Miller at Stony Brook University, and Allen Miller at the News Literacy Project, for their valuable insights into News Literacy as a curricular pursuit for all students—and to those who have been engaging in the structure of teaching and learning about news for decades and decades.

Finally, to the students who have come to the Academy for the last five years. This book would not exist if not for the energy, passion, and will that you bring each year to Salzburg, and that you sustain all year round in their local communities. It's all of you who drive us to keep building and growing and engaging and learning.

Paul Mihailidis, PhD
Assistant Professor, Emerson College, Boston, USA
Director, Salzburg Academy on Media and Global Change, Salzburg, Austria

Introduction -
News Literacy in the Dawn
of a Hypermedia Age

PAUL MIHAILIDIS

Emerson College, Boston, USA

Salzburg Global Seminar, Salzburg, Austria

MEDIA LITERACY, a concept born almost a century ago, has in recent decades come to the forefront of educational arenas dedicated to exploring the role of information in our lives. While many different definitions of media literacy exist for different disciplines, contexts, and uses, a common set of core assumptions include "the ability to access, analyze, evaluate, and communicate messages in a wide variety of forms (Aufderheide & Firestone, 1993). Since, definitions of media literacy have grown to include production, and incorporated digital technologies and semiotics into their purview (Gaines, 2010; Ofcomm, 2005; Potter, 2004; Silverblatt, 2001). While scholars have used the term media literacy to explore a wide range of different media functions—advertising, violence, pre-K learning, stereotypes, gender, and so on—few in the field have developed news as the focus of their scholarship.

News literacy, conceived under the umbrella of media literacy education, offers a new path towards addressing the possibilities and pitfalls that are created by the intersections where journalism, citizenship, and technology meet. It is an educational movement distinguished by the potential to re-energize a public increasingly distrustful of news media (Pew Research Center for People and the Press, 2009) and renew a demand for diverse, independent, credible, and deep civic information. News literacy acknowledges that in changing news environments, students of all ages need to learn about news not only through established practices and venues, but also as content pertains to new modes of voice, expression and perspective on a global scale.

Recent statistics show that around the world people are spending more time with mobile media technologies and social media platforms, and for more democratic purposes than ever before. Upwards of 700 billion minutes per month are now spent on Facebook worldwide, 88 billion Google searches per month are conducted globally, and twitter users are growing by the millions. As a result, more news outlets are exploring how these inherently collaborative and user-driven spaces are influencing their ability to report and disseminate news, and how best to serve a more active, expressive, and mobile public that is increasingly sharing information in integrated spaces of hypermedia activity.

News Literacy: Global Perspectives for the Newsroom and the Classroom is conceived to help prepare future media practitioners (and citizens) to embrace new media environments that can simultaneously empower both their craft and the civic voice. This means teaching not only about the various ways new technologies are used and to what end, but also how these tools can enable better reporting, more dialog with readers, and a more nuanced understanding of how information is processed through new media platforms. Such an approach addresses the increasing disconnect between new technologies, journalism, and civic participation that is increasingly apparent in communities across the globe.

This book gathers leading scholars, educators, and media makers from around the world to explore various new approaches to thinking about, examining, and evaluating news literacy and civic engagement around the following fundamental questions:

- What are the most pressing issues in news, media, and culture in a converged, digital and global media age?

- What are the best educational practices to help future media practitioners and citizens understand, engage and express information across borders, across cultures, and across divides?

The ideas, theories, and pedagogies that explore these questions employ somewhat diverse definitions of media literacy, news literacy, and related terms to help frame the context of their arguments. Historically, media literacy has been seen as a fluid term, applicable across disciplines, specialties, and pedagogies. It has incorporated health, advertising, politics, news, technology, gender, ethnicity, religion, and so on, under its umbrella of critical thinking and analysis of media and information. For this text, the idea of news literacy is seen as a subset of media literacy: the core concepts developed in the media literacy movement as applied directly to news. News, in this sense, adopts a traditional formulation of civic information about current affairs, and community issues relevant to awareness, engagement, and participation in local democratic processes. The result is a focus on how comprehension, evaluation, analysis, and production of news can help enable better teaching and learning strategies for more empowered, tolerant, aware, and active participants in 21st century civic democracy.

In certain chapters, for example, contributors will take time to develop their understanding of media literacy, news literacy, globalization, and of contemporary journalism, to help orient the reader to the ideas that they discuss, while in others the terms media literacy and news literacy are used interchangeably. Instead of attempting to draw fine lines that seem more limiting than elaborative, this book allows for definitional rigor and fluency to emerge in the pages of the book, and not be defined as something that limits the diverse and wide ranging ideas that follow. Resultantly, the diversity of definitions, ideas, and dispositions in this book will be utilized with a combination of theory, practice and pedagogy to help shape learning outcomes for understanding news, democracy, and civic voices in the 21st century.

The Changing News Landscape

Eleven years into the 21st century and the world has witnessed a sea change in the news industry. Traditional models for newspapers have eroded to their core, foreign bureaus have disappeared at alarmingly fast speeds, and journalists find their resources diminished and their ability to investigate a story at odds with the immediacy of the Internet. Television news, meanwhile, has become fertile ground for polarizing banter, editorial glamour, and self-serving sound bites. On the Internet, news outlets have found few sustainable models for news production and dissemination as they struggle to compete with a vast world of civic entrepreneurs. As communication industries across the board continue to deregulate, the news world has become a survival of the fittest, where ratings, markets, and profits trump content, depth, and diversity.

At the same time, new media technologies have provided an arena for information flow that is more collaborative, immediate, and more open than ever before. The result has been an information revolution in which mobile tools, social media platforms, and collaborative online spaces have changed basic habits of information production, dissemination, and reception. These collaborative networks have fundamentally shifted how individuals understand participation, expression, sharing, and community. Convergence of all media into one platform has also created integrated landscapes for citizens around the world (Benkler, 2006; Jenkins, 2006, 2009; Shirky, 2010). The result is an increasingly monitorial citizenry (Schudson, 1999)—surveying the vast expanse of information in short headlines and through links aggregated by content providers and curated by friends, only stopping to read more when motivated to do so. No longer are individuals so much seeking information than wading through the vast amounts of content—in print, video, and audio—that have become an integral part of their everyday media landscape. Twitter and Facebook are fast replacing newspapers and the television as primary means for information gathering and sharing. Physical community forums have been replaced by online spaces for dialog, discourse and collaboration.

This new media landscape has not changed the civic and democratic structures upon which our communities rest, but it has changed the way in which we participate. Writes media scholar Clay Shirky in *Cognitive Surplus* (2010): "The logic of digital media, on the other hand, allows the people formerly known as the audience to create value for one another every day" (52). The ability for citizens to share information and cultivate an active voice is now at the forefront of communities large and small. The result has been a more vibrant and diverse information landscape, but one with equal challenges for the continued survival and relevance of journalists in 21st century democracies.

Recent examples of technologies vast influence on news reinforce a new reality for journalism industries that are struggling to sustain their existence and integrity in a complex, fast-paced and borderless information environment. In 2008, US Senator Barack Obama leveraged social media to disseminate information, converse with journalists, and provide news updates for his constituents and supporters. This media strategy far surpassed any traditional press coverage in terms of reach and volume, cultivating a celebrity status rarely seen before in the political spectrum (Harfoush, 2009). In 2009, just after the elections in Tehran, Iran, concluded, protesters took to the streets to voice their displeasure with a dispute outcome. When traditional media were forbidden from telling these stories, young students took to Twitter to tell the world of the unrest and happenings in the aftermath of the riots (Gladwell, 2010). While the salience of these activities have been debated, they no doubt signal a clear expansion in the scope and scale of information sharing and civic voice in the face of oppression. In 2010, the non-profit news organization

Wikileaks, led by Australian dissident Julian Assange, published hundreds of thousands of documents, many of which were classified, detailing the United State's diplomatic strategies and foreign relations with allies and enemies around the world. These documents circulated globally in less than a day, viewed by millions. The Wikileaks phenomenon has forced public agencies and the public to reconsider the boundaries of secrecy, privacy and expression in a digital age.

More recently in China, the case of Noble Prize recipient and dissident Liu Xiaobo received little attention in Chinese mainstream media outlets, but social media enabled much domestic dialog, protest, and resistance against what is normally a very controlled and suppressed civic media spectrum. After calls for his immediate release from prison, foreign and Chinese social media sites began extensive dialog and protest in support of Xiaobo. While the Chinese Government resisted calls for his release, they could not silence the anonymous protesters, or the global reach of their voices.

Now in 2011, the world is witnessing what many are calling the largest civic uprising largely enabled by social media platforms and mobile technologies. The Arab Spring, instigated on December 19, 2010, when jobless graduate Mohamed Bouazizi, whose fruit cart was seized by authorities, set himself on fire in the Tunisia city Sidi Bouzid. This act touched off protests in the relatively peaceful Tunisia that eventually spread throughout the entire region. Citizens, who rarely had the ability or capacity to organize, now had a slew of tools to confront Autocratic regimes, outside the bounds of traditional mass media. These tools—Facebook, Twitter, Ushahidi, and the like—were mostly conceived as leisurely social outlets for connecting with friends and sharing personal information. The Arab Spring, and those events before and after, have signified a distinct tipping point for the use of such tools as participatory instigators and viable news outlets for journalists and audiences.

All of these examples are predicated on more citizens turning towards online spaces for news before traditional forms of content delivery (Pew Internet and American Life Project, 2010). And not only are they turning to these spaces to receive news, they are turning to these spaces to share, express, collaborate, and *act* around news, creating a new dynamic for the relationship between news, technology, and democracy: the implications of which are vast for global citizenries in the 21st century.

The new technologies that have impeded on traditional news models have presented many challenges for news organizations. Most notably, fears of a "dying" journalism field revolve around the loss of credibility, investigation, objectivity, and in-depth news being provided for the public. While these fears are indeed warranted, the opportunities for new dynamics in news and journalism are growing. Writes Alex Jones in *Losing the News* (2009):

> A case can be made that the core [of the news industry] will not only survive, but grow more weighty through new forms of news media...traditional media are trying to find new ways to report news that will appeal to a younger, Web-savvy audience, and creating new publications and Web sites in response to reader tastes. (21)

With every shift in technology comes a shift in information. While the merits of these shifts will be debated for some time to come, the new reality for news and journalism will have implications on the future of democratic societies in the 21st century. Indeed, as audiences become more Web-savvy, and demand information that is mobile, tailored to social networks, and immediate, news organizations will be forced to respond. Four major innovations in digital media are currently reforming information flow and influencing the future of journalism.

1. *New Civic Voices* – Social media tools and platforms have not only allowed for more efficient group collaboration and community dialog, but have also forced traditional media organizations to create spaces premised on civic voices. Beyond simple comments at the end of articles and multimedia packages, citizens now have dedicated spaces for reporting (iReports, Wikinews), the ability to create and share content, and the tools to spread this information effectively to their intended audiences. These voices have become the public sphere of the Internet, a virtual coffee shop of ideas that are shared, remixed, and repurposed to create a diverse center of opinion, judgment, and commentary on any and all issues, events, and topics.

2. *Mobile Technologies* – The new technological evolution has been predicated on the replacement of time and space with immediacy and convenience. Mobile technologies, and the digital media tools they foster, have created an environment for information in which the citizen receives, shares, publishes, reports, and collaborates often without need-

ing to been in a certain place at a certain time with a certain infrastructure. Nor do they need much formal training to learn how to use these technologies, and share information. The result has been the replacement of investigative journalism, and on-the-scene reporting, with journalists curating mountains of information created by citizens using mobile technologies to capture and share information to a wide audience in real time.

3. *Participatory Tools* – Perhaps the Internet's greatest influence on news and journalism is that its existence is premised on participation. Digital spaces fostered by the Internet are dependent entirely on collaboration and civic participation. This has changed traditional top down models of information dissemination, and largely eliminated what Clay Shirky (2008) calls the transaction costs of traditional media production and consumption. When participation is the first instinct of the user, a natural competition is created for information that is endlessly created and made public. Instead of carving out their own space for "news," journalism organizations have had to work within converged landscapes to provide meaningful information that users demand, often in the spaces predominately occupied by the users themselves.

4. *Spreadability* – Finally, the ease by which information spreads have diminished the boundaries for information. Information today spreads far, wide, and fast. The combination of new technologies and the urge for citizens to share and publication information has resulted in the vast spread of information around the world. The result has been a shift to either hyperlocal content, or content that will travel beyond cultures, divides, and specific audiences. While the ease of dissemination has increased the opportunity for journalism to have a greater impact, it has also created new competing factors and the loss of a specific audience.

The model below re-creates this new landscape for journalism to include the four new realities of news listed above, and shows them in the context of the increasingly shared spaces occupied by journalists and citizens.

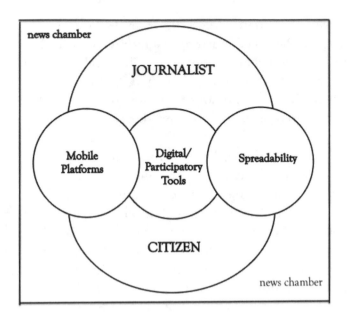

Figure x.1: Concentric Model for 21ˢᵗ Century News. [Courtesy of Author]

   The model premises a need to explore the new realities of the relationship between the journalist and the public, and how the new space is being negotiated by individuals, communities, and nations. These new spaces increasingly co-inhabited by journalists and citizens will largely dictate the future of journalism, media, and citizenship in the 21ˢᵗ century. As a result, there exists a growing need to educate youth and adults about news around new modes of communication and media technologies in an increasingly global media landscape. The news literacy[1] educational movement is premised on exploring how to best prepare journalists and citizens for lives of active inquiry and participatory citizenship in information societies worldwide.

---

1 In certain chapters of *News Literacy* the terms news literacy and media literacy interchange and/or are conflated. Two reasons support the interchangeable nature of the terms in this text: News Literacy is a subset of Media Literacy, and thus is premised on the foundational scholarship and curricular efforts associated with "media" literacy, and; there is a dearth of literature on news literacy, and so chapters develop their arguments through media literacy scholarship.

Premise: The Salzburg Academy on Media and Global Change

The idea for this book originated at the *Salzburg Academy on Media and Global Change*. With a network of over 250 students and 50 faculty from over 25 countries worldwide, the Salzburg Academy has transformed global media education by bringing together over 50 students and 10 faculty each summer to Salzburg, Austria, from all over the world, and charging them to build products that characterize media and citizenship as inherently global, and representative of the cross-cultural media environments now occupied by a majority of individuals worldwide. Through an interdisciplinary and cross-cultural media literacy approach, the Academy has developed a new framework to teach students not only to think critically about media and media messages, but also to defend and appreciate the necessity of free and diverse media systems for free and diverse global communities.

In 2007, fifty-two students from fourteen countries over five continents gathered at the Schloss Leopoldskron in Salzburg, Austria, for three weeks to create educational content around media, freedom of expression, democracy, and citizenship. The premise of this gathering—the inaugural Salzburg Academy on Media and Global Change—was that a truly global collaborative effort is a prerequisite to creating a truly global media literacy educational experience.

In the five years since its birth, the primary outcome of the Salzburg Academy, in addition to the individual growth and transformation documented by those who participated, was a student-created curriculum on global media literacy. Students, in interdisciplinary groups, work to create a series of case study-driven learning modules. The learning modules include an introductory case study, a set of classroom and homework exercises, discussion questions modeled along the 5 A's of media literacy,[2] multimedia stories on the topic at hand, and additional resources for further exploration. The Academy students

---

2 The 5 A's of media literacy framework, developed in year one of the Academy, and authored by the editor of this book, offers a unique foundation for exploring how media can bridge cultural, political, and ideological divides. Students who are able to engage with global media around the 5 A's—*access* to media, *awareness* of media's power, *assessment* of how media cover international and supranational events and issues, *appreciation* for media's role in creating civil societies, and *action* to encourage better communication across cultural, social, and political divides—can develop habits of inquiry around how media defines issues on a global scale. The "5 A's" enable a continuum starting with an understanding that there is no democratic society without access to information, and concluding with the idea that in today's hypermedia environment, we all have the ability to be active participants in global communities.

work in diverse groups to collectively build, edit, and finalize each lesson plan, making sure the scope of each product is global in scope, collaborative in its development, and adaptable for diverse audiences.

The Academy's global media literacy curriculum reinforces critical inquiry and analytic skills with modules that emphasized the vital importance of free and independent media in building and supporting civil society (Mihailidis, 2009). Through a collaborative, ground-up approach to teaching and learning about global media, the Academy has created not only dynamic education products, but also a core group of future media practitioners that have gained invaluable insight into how media systems define cultures and identities in foreign nations. This involves thinking beyond borders, and beyond specific media, to understand the unique ways media defines civil society across the globe.

The chapters in this book are written by scholars who have participated as faculty in the Salzburg Academy over the last five years. Their work here stems from the dynamic engagement, dialog, and learning that happen each summer in Salzburg. The challenge of a book like this is how to best appropriate the diverse expertise and background of the faculty into a coherent and cohesive text. While the book is organized in two parts—one on theory and one on pedagogy—readers will notice that some chapters fit neatly into this division, while others straddle the line between the two, and still a few diverge at points entirely. This is not meant to confuse the organization of the text, but rather to find a way talk about news literacy that is entirely integrated into the fabric of the contributing scholars and the Academy itself.

For example, in Part Two of the book, Chapters 5 – 7 depart rather directly from Chapters 1 – 4, and further Chapter 8 departs in some respects from Part Two and other chapters in that section. The conclusion, furthermore, is told through a working example that exemplifies the civic empowerment the Salzburg Academy attempts to cultivate in its participants. The organization of this book, in the context of the Salzburg Academy, is meant to both connect the various theoretical and pedagogical considerations for news and media literacy in global contexts, and at the same time allow for the diverse voices and ideas that permeate truly global settings to be heard.

Overview: Bridging Theory, Practice, and Pedagogy

Responses to the increasingly converging and mobile technologies for journalists, and the increasingly blurred line between reporter and opinion-giver, are perhaps now more vital than ever before. It is within this landscape that news literacy can serve as the bridge for building more dynamic, relevant, and comprehensive approaches to teaching, learning, and understanding news across the globe.

Part I of *News Literacy* explores critical theory and scholarly inquiry around news literacy, media education and journalism in digital and global contexts. If future journalists are to be effective in reporting information with tools that lend to more immediate, global, and multi-platform information, the various aspects of these new realities must be seen in light of their various implications for 21st century society.

As civic media grow, individuals play increasingly central roles in interdependent and interconnected news flow and information sharing. Such new dynamics are democratizing participation and observation that allow journalists to gather footage that is raw, immediate, and intensely subjective, says Bournemouth University professor Stuart Allan. Allan takes a historical orientation to the development of media and news literacy over time, and considers the extent to which the "citizen" can help forge new points of connection, possibly even emphatic engagement between *us* and *them* (the citizen and the journalist). Chapter One places news literacy at the center of the re-inflection, conceived to empower the citizen as the primary source for news now devoid of time, space, and patience. Using the recent civil unrest in Greece as his entry point to the present, Allan develops a nuanced understanding of how social media platforms enable a new relationship between the public, the media, and the government, predicated on active, collaborative, and collective civic voices The relationship that results from this new engagement is seen as the crux of networked power and global connection in an increasingly shrinking and connected globe.

The journalist's ability to report in digital media spaces is largely dependent on models that support new forms of information production, dissemination, and reception. In Chapter Two, Iberoamericana University professor Manuel Guerrero develops a model of a media literate *prodience*, which is premised on the connection of news content and production in open society. This model of relationships outlines production factors and information content across levels developed by Shoemaker and Reese (1996) (individual, corporate, extramedia and ideological) to redefine the role of news flow in media environments that are premised on continual, anonymous, and open sharing and repurposing of news by individuals, organizations, and political bodies.

The new model proposed by Guerrero incorporates a greater reliance on civic competencies to build accountability in 21st century news industries: the result of which is a call for news/media literacy to bridge the disintegrating divide between the news producer and audience.

If news literacy is the bridge between the producer and receiver, what are the ways in which this pedagogical movement can create more active and reflexive global consumers of information? University of Texas professor Stephen Reese picks up in Chapter Three where Guerrero left off to ponder pedagogical models that successfully respond to the more complex, global, and multi-layered news and information systems born largely from technological advancements over the last few decades. Media literacy has long advocated critical thinking as its core approach to engaging with information of all kinds. With the relative explosion of alternative sources and diverse perspectives leveraged by access to technologies that are global and open, new approaches to training journalists for productive careers hinge on media literacy, and its subset news literacy, Reese notes. Through an exhaustive review of pedagogical initiatives in recent years, Reese concludes that the need to engage aspiring students with deep reflection about their role as citizens in information societies, and how their information habits are predictors for the larger flow of global news.

The last chapter in part one combines the ideas developed in the first three chapters around news production, dissemination, technologies, and pedagogy, to explore the continued and vital necessity of the practicing journalist for democracy in information societies. If new avenues for information exchange are putting traditional news models to the test, Catholic University of Argentina professor and journalist Raquel San Martín posits, how will digital media platforms enable new values and representations of news practices. Banter about the "death" of journalism is increasingly common in countries around the world, with few exceptions. Instead of lamenting the changes as losses, San Martín develops a new ecosystem for journalism, one that mixes professional with amateur, journalist with citizen. The new models posited in this chapter reposition the practitioner as a hyperlocal, community driven reporter who leverages information across borders, cultures, and communities through new tools and technologies that can serve larger audiences. The chapter develops news literacy as the bridge between the practitioner and the citizen, one that allows them to co-depend on each other for dynamic, investigative, and diverse information flow to local, national, and global communities.

Part II of *News Literacy* builds on Part I to explore pedagogical approaches for teaching and learning about news and journalism. Chapters in Part II emphasize concrete strategies and practices for formal and informal education about news and journalism in digital and global contexts. Educational approaches highlighted include: comparative research, multimedia production, case study construction, and democratic development through news literacy. Connecting theoretical insight to pedagogical practice is essential for media practice in the 21st century. For as the lines between citizen and journalist become indistinguishable, how these civic roles are understood must be revisited in the context of new pedagogical approaches.

An essential part of teaching and learning about media in the 21st century hinges on creating shared dialog around issues, topics, or events, argues Catholic University of Chile scholar Constanza Mujica. The case study approach can enable critical engagement with media issues through specific examples that can cross borders, cultures and divides. In Chapter Five Mujica applies the foundations of media literacy—comprehension, analysis, evaluation, production—to highlight approaches to teaching with the case study platform. The 5A's of media literacy framework (Mihailidis, 2009) is presented as a way to help connect theory to practice in the case study methodology.

In today's era of digitized, globalized media, students must not only be critical consumers of information, but also critical producers of messages, write Florida International University professor Moses Shumow and University of Miami professor and vice dean Sanjeev Chatterjee. With the barriers to producing and publishing reduced to internet access and a login, the journalists and citizens of today will be collaborating, creating, and publishing more than ever before. News literacy emphasizes critical thinking and savvy information navigation as essential to fostering diverse democratic dialog. While analysis and reflection represent a large part of this equation, Shumow and Chatterjee propose pedagogical structures focused on innovative approaches to learning about global news through multimedia production. Their chapter highlights a series of examples of production based education initiatives that resulted in the creation of innovative and unique forms of storytelling. They develop ideas for classroom environments that use news literacy and multimedia production to better prepare future producers and consumers of information in digital spaces.

More recently, in addition to case studies and multimedia production, comparative analysis has attracted the interest of media literacy scholarship largely due to the acceleration of globalization and the parallel expansion of the Internet that brought once distant and inaccessible media content to the fingertips of almost anyone in the world. This new reality has had significant influence on how news is reported in both local and global communities. American University of Beirut professor Jad Melki emphasizes the need for established frameworks of comparative analysis of global news as essential to news literacy education in the 21st century. The focus of teaching and learning comparative frames and agendas rest largely on shifting pedagogical approaches from theoretical to practical, allowing students to engage with the content in a manner that reflects their media habits and aspirations. Melki offers worked examples of comparative analysis, and hands-on guidelines for educators and scholars interested in pursuing this method of inquiry further.

The pedagogical approaches laid out in Part II of this text together can work to deepen democratic discourse, argues Ugandan media analyst George Lugalambi. Taking the case of media development in Uganda, Lugalambi connects news literacy to the sustainability of flourishing civic democracies. For, as he writes, "without informed citizens, goes the argument, it is impossible for the population to hold leaders and governments as a whole accountable." With the growing presence of new and mobile media technologies in Uganda and throughout Africa, how youth are taught about the opportunities these new digital tools have for civic dialog is increasingly important to the quality of democratic growth. Lugalambi supports his argument with a case study on Africa through the pedagogical approaches of media and news literacy, highlighting approaches in the previous chapters and connection education to civic-oriented outcomes.

Ultimately this book, calls for news literacy to help in cultivating active and engaged citizens and journalists and to create cross-national dialog for good governance and civic participation. Building on the central themes running throughout the text, this call to action will conceptualize news literacy as the core component for building engaged, informed and active citizens that can play more inclusive roles in their social and civic processes. In the conclusion, University of Maryland professor and media scholar Susan Moeller argues that these core competencies are also closer to journalism than at any point in the past, and that news literacy, as conceived in these chapters, can help to strengthen journalism across the globe. The conclusion will highlight theories of participatory media, citizen journalism and media literacy education to create an active approach to building strong platforms for news literacy education in the 21st century.

News literacy has the potential to bridge the increasingly indistinguishable divide between the reporter and the audience, the journalist and the citizen. For while the role of the journalist in democracy is perhaps now more important than ever before, harness the technological advancements of the digital media age to offer more diverse, collaborative, and wide reaching information has often lagged in the educational arena. Scholars, educators, practitioners, and policy makers all have vested interest in understanding how these relationships manifest themselves in the 21st century. This book aims to build a cohesive and in-depth look at how newsrooms and classrooms can benefit from harness the power of new media tools for more vibrant media industries large and small, and more active and engaged citizens worldwide.

REFERENCES

Aufderheide, P., and Firestone, C. (1993). *Media Literacy: A Report of the National Leadership Conference on Media Literacy.* Cambridge, UK: Polity Press.

Benkler, J. (2006). *The Wealth of Networks: How Social Production Transforms Markets and Freedom.* New Haven, CT: Yale University Press.

Gaines, E. (2010). *Media Literacy and Semiotics.* New York: Palgrave.

Gladwell, M. (2010). "Small Change: Why the Revolution Will Not Be Tweeted." *The New Yorker Magazine,* 4 October 2010.

Harhoush, R. (2009). *Yes We Did: An Inside Look at How Social Media Built the Obama Brand.* Berkley, CA: New Riders Press.

Jenkins, H. (2006). *Convergence Culture: Where Old and New Media Collide.* New York: NYU Press.

Jenkins, H., Purushotma, R., Weigel, M., Clinton, K., and Robinson, A. J. (2009) "Confronting the Challenges of Participatory Culture: Media Education for the 21st Century." A Report for the MacArthur Foundation, Washington, DC.

Jones, A. (2009). *Losing the News.* Oxford: Oxford University Press.

Mihailidis, P. (2009). "Beyond Cynicism: Media Literacy and Civic Learning Outcomes in Higher Education." *International Journal of Learning and Media,* 1(3), pp. 19-31.

OFCOM. (2005). "Media Literacy Audit. Report on Media Literacy Amongst Children." Office of Communication. London. Retrieved on December 2, 2010: http://www.ofcom.org.uk/advice/media_literacy/medlitpub/medlitpubrss/children/children.pdf

Pew Internet and American Life Project. (2010). "Understanding the Participatory News Consumer," by Kristen Purcell, Lee Rainie, Amy Mitchell, Tom Rosenstiel, Kenny Olmstead. March 1, 2010. Retrieved on October 10, 2010: http://www.pewinternet.org/Reports/2010/Online-News.aspx

Pew Research Center for People and the Press. (2009). "Press Accuracy Rating Hits a Two Decade Low," by Andrew Kohut. September 13, 2009. Retrieved December 4, 2010: http://people-press.org/report/543/

Potter, J.W. (2004). *Theory of Media Literacy. A Cognitive Approach.* Thousand Oaks, CA: Sage Press.

Schudson, M. (1999). "Good Citizens and Bad History: Today's Political Ideals in Historical Perspective." Paper presented at conference on *"The Transformation of Civic Life,"* Middle Tennessee State University. November 12-13. Retrieved October 18, 2007, http://www.mtsu.edu/~seig/paper_m_schudson.html

Shirky, C. (2010). *Cognitive Surplus: Creativity and Generosity in a Connected Age*. New York: Penguin.

——. (2008). *Here Comes Everybody: The Power of Organizing without Organizations*. New York: Penguin.

Shoemaker, P., and Reese, S. (1996). *Mediating the Message: Theories of Influence on Mass Media Content*. New York: Longman.

Silverblatt, A. (2001). *Media Literacy: Keys to Interpreting Media Messages*. 2nd Ed. Westport, CT: Praeger.

Part One

Theoretical Models for News Literacy Education

Chapter 1 -
Civic Voices: Social Media and Political Protest

STUART ALLAN

Bournemouth University, England

'...the popular Press is the most powerful and pervasive
de-educator of the public mind...'
F.R. Leavis and Denys Thompson (1933, 138)

'Let's keep this to our original reporting, information we collect our-
selves and let's leave outside what the media establishment says'
Indymedia.org contributor during student protests in Greece,
December 2008

Introduction

There is little doubt that the significance of events at the heart of what is currently being described as the Arab Spring of 2011 will be defined by outcomes
we can only begin to anticipate today. In following journalistic accounts of
what is transpiring, it is intriguing to note the extent to which young people's
use of internet and social media – such as Facebook, Flickr, Twitter, YouTube
and the like – have proven to be newsworthy topics in their own right within
the coverage. A *New York Times* article, 'A Tunisian-Egyptian link that shook
Arab history,' underscores this point:

> As protesters in Tahrir Square faced off against pro-government forces, they drew a
> lesson from their counterparts in Tunisia: "Advice to the youth of Egypt: Put vinegar
> or onion under your scarf for tear gas."

The exchange on Facebook was part of a remarkable two-year collaboration that has given birth to a new force in the Arab world — a pan-Arab youth

movement dedicated to spreading democracy in a region without it. Young Egyptian and Tunisian activists brainstormed on the use of technology to evade surveillance, commiserated about torture and traded practical tips on how to stand up to rubber bullets and organize barricades [...] (*The New York Times*, February 13, 2011).

The article proceeds to explain that the young protestors have been 'breaking free from older veterans of the Arab political opposition' over recent years so as to form an Egyptian youth movement intent on challenging state corruption and abuse. Informal online networks, using a Facebook group as their nexus, have succeeded in setting in motion a range of tactics to articulate nonviolent resistance. While their relative success is impossible to determine at this point in time, one may be forgiven a certain cautious optimism that virtual civic spheres enlivened by public participation, deliberation and engagement are currently emerging with the potential to empower ordinary people to renew their efforts to extend democratic change and human rights.

In assessing the specific features of the uprisings unfolding across the Middle East, it is important not to overlook the fact that these forms of citizen protest have been years in the making. Bold, even at times triumphalist claims about digital technologies have featured prominently in impassioned discussion across the mediascape, where cyber-enthusiasts herald them as tools of liberation responsible for ushering in near-instant revolutionary change. Contrary voices tend to be sharply dismissive, insisting that conventional types of political mobilisation and protest are being overlooked in the hype. A more measured appraisal, situated between these two polarities, would necessarily recognise the structural imperatives underpinning these ostensibly spontaneous eruptions of dissent. Important here, I would suggest, is the need to look beyond otherwise starkly rendered assertions in order to investigate the lived, complex – and frequently contradictory – forces giving shape to collective re-imaginings of civic cultures. This is particularly the case where young people's identity politics are concerned. Social networks facilitated by digital technologies help to engender the conditions whereby they learn to become citizens willing and able to engage in the world around them.

To contend that citizenship today demands media literacy is to invite new thinking about the aims and purposes of media education. At a time when some commentators are posing awkward questions about precisely what, if anything, is 'new' about new media, others are wondering aloud about whether the end of media studies – traditionally defined – is now in sight. In the view of the latter, the rapid convergence of 'old' and 'new' media and their attendant technologies necessitates a recasting of what should constitute media pedagogy so as to more fully embrace the brave new world of 'Web 2.0'. Regardless of one's stance on these and related issues, it seems to me that ensuing discussions should be sufficiently sensitive to the multiple inflections of 'media literacy' in such contexts. The need for this sensitivity becomes particularly apparent, I would suggest, once the everyday uses of media by young people are centred for investigation. The learning dynamics given expression in their creative, frequently innovative re-appropriation of 'old' media forms and practices in emergent digital environments render problematic certain longstanding conceptual categories of media studies (see also Buckingham, 2003; Carter and Allan, 2005; Dahlgren, 2007; Guedes Bailey et al., 2008; Lister et al., 2008; Loader, 2007; Messenger-Davies, 2010; Mihailidis, 2009). More than that, however, familiar assumptions about what should constitute 'literacy' in the first place are being increasingly called into question.

In seeking to contribute to discussions about global news literacy from this vantage point, this chapter begins by considering a formative intervention in media education from the 1930s, namely F.R. Leavis and Denys Thompson's attempt to explain why educators must 'systematically inculcate' in young people a 'critical habit' when negotiating the influence of advertising, cinema, the popular press, and the like, so as to ensure they strive to uphold 'positive standards' in their daily lives. While at first sight this highly polemical treatment may seem curiously anachronistic, I shall endeavour to show that it helps us to discern tensions that continue to resonate in current debates about news literacy. In the remainder of this chapter's discussion, these tensions are explored in conceptual terms, as well as with respect to recent examples of young people's active negotiation of emergent forms of digital citizenship. On this basis, it will be argued that research into news literacy must rethink the current emphasis placed on the social competencies associated with news consumption in order to attend to the ways in which young people are actively recrafting social media as resources in the service of elaborating new, empowering forms of civic engagement in public life.

Critical Awareness

While general claims made about the origins of media education in Britain tend to prove contentious when scrutinised closely, few would dispute that the publication of F.R. Leavis and Denys Thompson's book *Culture and Environ-*

ment in 1933 was a noteworthy moment in this history. Their savage indict-
ment of what they perceived to be the media's degenerative influences, particularly where the corruption of the morality of young people was concerned,
pinpointed a host of fears about the cultural decline of society. According to
Leavis and Thompson (1933):

> Those who in school are offered (perhaps) the beginnings of education in taste are
> exposed, out of school, to the competing exploitation of the cheapest emotional re-
> sponses; films, newspapers, publicity in all its forms, commercially-catered fiction – all
> offer satisfaction at the lowest level, and inculcate the choosing of the most immedi-
> ate pleasures, got with the least effort. (3)

As they proceed to make clear, responsibility for instilling in young people the
means to uphold 'civilising values' would have to be borne by teachers of English literary criticism. Young people needed to be equipped with the means to
distinguish between 'authentic' and 'inauthentic' culture as a matter of 'taste',
Leavis and Thompson (1933) argued, if the 'human spirit' was to be sustained.
In their words:

> ...if one is to believe in education at all one must believe that something worth doing
> can be done. And if one is to believe in anything one must believe in education. The
> moral for the educator is to be more ambitious: the training of literary taste must be
> supplemented by something more....[W]e are committed to more consciousness; that
> way, if any, lies salvation. We cannot, as we might in a healthy state of culture, leave
> the citizen to be formed unconsciously by his [sic] environment; if anything like a
> worthy idea of satisfactory living is to be saved, he must be trained to discriminate and
> to resist. (4-5)

The need to engage with media forms and practices in the classroom was
therefore an exigent priority, they insisted, not least because the future of 'a
world of depressed and cynical aimlessness' stood in the balance. The teaching
of media criticism, if not quite respectable in their eyes, was nevertheless becoming increasingly inevitable. Once a student is provided with an awareness
of adequate 'standards', Leavis and Thompson maintained, 'the offerings of
the mass media will appear cut down to size.'

While journalism receives only passing consideration, the popular press –
as the quotation at the outset of this chapter indicates – is singled out for
scathing condemnation. Of particular concern is the 'distraction' it engenders
(and here the authors wonder whether 'dissipation' may be a better word),
with the reader implicitly chastised for being lulled into a stupor of sorts,
whereby insufficient attention is focused on events. 'In the popular newspaper

the tendency of the modern environment to discourage all but the most shallow and immediate interests, the most superficial, automatic and cheap mental and emotional responses, is exhibited at perhaps its most disastrous,' Leavis and Thompson (1933, 102) write. Given the primary purpose of the newspaper to effectively inform members of the public about the world around them, the emphasis placed on generating amusing distractions amounts to an 'adulteration of function' in their view. 'It is well to insist,' they continue, 'that even more mischievous than misrepresentation and the exploitation of ignorance and the whipping up of herd-prejudice is the habit of irresponsibility about public matters that the Press fosters' (103). Much of the ensuing discussion in the book's remaining pages is devoted to considering various 'exercises in observation and comparison' intended to help break this 'habit of irresponsibility' in the interest of the wider community.

One can only speculate about what Leavis and Thompson would think about media studies as it is taught today, of course, but it is readily apparent that that the arguments they advanced reverberate in debates about media literacy. If the participants in current discussions tend to be less inclined to use a language of 'taste' and 'discrimination' to express their beliefs, many of them nevertheless continue to sound the alarm over what they perceive to be the damaging 'effects' of media representations on young people. The rationale for media education, as a result, is often couched in the terms of first identifying and then resisting the pull of harmful influences. A media-literate student, it follows, is one who has been sensitised to the ways in which various media institutions seek to 'distort' or 'falsify' reality, usually in accordance with commercial interests, in order to further the cause of consciousness raising. As such, she or he needs to be sufficiently inspired to uphold higher values of decency, that is, a strong sense of moral purpose. Underlying corresponding notions of media literacy, then, is the assumption that media education needs to operate at the level of inoculation. That is to say, the student must be partly exposed to the debilitating forms of media influence in the classroom so as to ultimately enhance their immunity from manipulation. In this way, the teacher fulfils their role of ensuring the 'improvement' of the student's personal sense of refinement and her or his capacity for cultivated judgement.

It goes without saying, of course, that prefigured in any approach to media pedagogy is an array of presuppositions about how students actually learn, and – equally crucial – why they want and need to learn in the first place. Approaches striving to reaffirm the tenets of Leavis and Thompson's guidelines risk taking for granted the belief that students are willing to be instilled with the desire to resist the enticements of media texts and imagery. In other words, they assume that students can learn to share their teacher's 'finer sensibilities', and that once so qualified will feel compelled to elevate themselves from the realm of 'low' or 'popular' culture into that of 'high' culture or the 'fine arts'. No doubt there are contexts where such an approach might be deemed relevant – the teaching of aesthetics in art history, for example, may be one – but there is little chance that this will prove to be the case in the media studies classroom. Far from regarding the media as pernicious purveyors of 'cheap thrills' or feeble-minded 'distractions', today's student is much more likely to consider different genres of media content as sources of intense self-identification, and all the more pleasurable for it. Moreover, many such students are passionately committed to securing their own forms of media-making, finding in the tools of do-it-yourself composition and critique the means to articulate their own perspectives in a manner that recurrently calls into question 'proper,' 'appropriate' or 'legitimate' value predispositions.

A Multiplicity of Literacies

While young people's interest in media-making is hardly a new development, much of the media literacy literature revolves around the evaluative critique of mainstream media forms and practices as a worthy end in its own right. Interesting to observe in this regard, however, is the extent to which various early considerations of how 'old media' ideas about literacy were being recast by interactive technologies highlighted the limitations of this text-centred emphasis from the start.

Douglas Kellner (2002) argued almost ten years ago, for example, that we need to develop 'multiple literacies' so as to better respond to the globalising demands for a more informed, participatory and active citizenry in political, economic and cultural terms. Literacy, in this conception, 'comprises gaining competencies in effectively using socially constructed forms of communication and representation' (92). More specifically, media literacy 'helps people to use media intelligently, to discriminate and evaluate media content, to dissect media forms critically, and to investigate media effects and uses' (Kellner, 2002, 93). In the new multimedia environment, Kellner (2002) maintained, this type of literacy had never been more important, especially with regard to the development of skills to create 'good citizens' motivated to play an active role in social life. He pointed out that the same technologies of communication capable of turning 'spectators into cultural zombies' may, at the same time, be used to invigorate democratic debate and participation. The problem, then as now, was how to bring about the latter on the terrain of the former. That is, how to take seriously the texts of popular culture enjoyed by students, recognising and respecting their ideas, values and competencies, without 'romanticiz[ing] student views that may be superficial, mistaken, uninformed and full of various problematical biases' (94). One way forward, Kellner suggested, is to adapt new computer technologies to education so as to facilitate the development of new literacies.

In seeking to expand upon familiar conceptions of literacy, Kellner drew attention to emergent forms of what he terms 'computer literacy'. Important here, he argued, is the need to push this concept beyond its usual meaning, namely as the technical ability to use computer programs and hardware. A broader definition, it followed, would attend to information and multimedia literacy as well. That is to say, Kellner's extended conception of computer literacy would include learning how to use computers, locate information via search engines, operate e-mail and list servers, and construct websites. Computer and information literacies, Kellner (2002) wrote, involve 'learning where information is found, how to access it, and how to organize, interpret and evaluate it' (95). At the same time, they also entail 'learning how to read hypertexts, to traverse the ever-changing fields of cyberculture, and to participate in a digital and interactive multimedia culture that encompasses work, education, politics, culture and everyday life' (Kellner, 2002, 95; see also Allan, 2002; Hassan, 1999; Lievrouw and Livingstone, 2002; Sefton-Green, 1998; Warnick, 2002). Clearly at stake here, then, is the teaching of more than just technical forms of knowledge and skills. By stretching the notion of literacy to include new strategies of reading, writing, researching and communicating abilities appropriate to a larger 'computer culture,' Kellner was helping to discern the conceptual space necessary to engage with an array of different, yet interrelated, types of information processing that possessed the potential to open up opportunities for alterative types of media practice to emerge.

Reading this and related research with the benefit of hindsight, one recognises the extent to which early discourses of 'computer literacy' inform certain formulations of media literacy today, namely those striving to interweave relevant aspects of 'news literacy,' 'information literacy,' 'visual literacy,' and 'digital literacy' to advantage. The scholarship devoted to these and related conceptions of literacy is voluminous, and need not be rehearsed here. Rather, it is sufficient to note for our purposes the growing awareness amongst re-searchers and practitioners of the reasons why 'literacy' must necessarily stretch to encompass creative forms of making, sharing and collaborating in a manner alert to a community's – lived or virtual – collective priorities. 'When people have digital and media literacy competencies, they recognize personal, corporate and political agendas,' Renee Hobbs (2010) observes, 'and are em-powered to speak out on behalf of the missing voices and omitted perspectives in our communities.' Moreover, she adds, by 'identifying and attempting to solve problems, people use their powerful voices and their rights under the law to improve the world around them' (17). While as a broad assertion this risks sounding a little idealistic, Hobbs is usefully underscoring – at least in my reading – the importance of discerning how and why such competencies be-come socially relevant, and thereby politicised in normative terms.

In danger of being glossed over in discussions of young people's skills in negotiating the affordances and constraints of digital technologies, I would suggest, is the extent to which a politics of citizenship informs their participa-tory cultures. Such a view, I readily acknowledge, challenges the oft-rendered assertion that young people are apathetic, even cynical about the prospect of active involvement in their communities. What may appear to be passive dis-engagement, however, is most certainly political nonetheless (indeed, all the more so because it is seldom recognised as such). Where 'politics' is allowed to be defined narrowly within a discourse of partisanship, it is safely contained within the domain of voting and political parties; this when young people's everyday negotiations of social hierarchies, divisions and exclusions will be much more likely to be considered by them to be relevant to their lives. It fol-lows that sweeping claims regarding their apparent disaffection are open to challenge on these grounds, yet it is telling how often their use of digital media – especially where social networking is concerned – is singled out for atten-tion, if not moral censure.

Amy Mitchell's (2010) research contends, for example, that '[s]ocial media tools and mobile connectivity provide citizens with a deeper and more direct relationship with the news,' one which suggests that 'news grazers' are any-thing but 'aimless wanderers.' Mobile devices not only bring new information sources into the mix, she maintains, they dramatically alter what happens to news reports after they appear. Young people, in particular, will be inclined to search, filter, react to, and share news they consider interesting with one an-other, often relying on social networking to serve as a 'personal editor' of sorts helping to 'determine their front page information' (27) [see also Pew, 2011]. Mizuko Ito (2010), in describing the apparent 'generational gap' in how people of different ages regard social media, points out that disagreements over 'what participation in public life means' are often at issue. 'Whether it is teachers trying to manage texting in the classroom, parents attempting to set limits on screen time, or journalists painting pictures of a generation of networked kids who lack any attention span,' she writes, 'adults seem to want to hold on to their negative views of teen's engagement with social media' (18). Given that young people deliberately sidestep 'institutional gatekeepers' wherever possible, she adds, literacy becomes a 'byproduct' of social engagement. Larry Rosen (2010) reaffirms this observation, maintaining that as 'the pace of technological change accelerates, mini-generations are defined by their distinctive patterns of media use, levels of multitasking, and preferred methods of communication' (25). Efforts to identify the 'distinctive digital habits of mini-generations,' it follows, need to be aware of differences 'in their values as well as levels of social and political activism' (25-26). Precisely what counts as political activism varies considerably across these mini-generations, of course, but several researchers have sought to highlight the ways in which social networking is recasting familiar assumptions about civic participation.

One of the first instances where the political implications of these emergent forms of connectivity attracted media attention in their own right occurred in Greece in December, 2008. In the next section, our attention turns to consider the ways in which student protestors demonstrating against state institutions, not least the police, called upon social media to shape the articulation – and co-ordination – of their dissent. Moreover, it will be shown that equally important for rethinking conceptions of news literacy is the recognition that young people's contributions to real-time reportage of violence on the streets are changing the norms, values and priorities of journalism itself.

Voices of Protest

'Rebellion is deeply embedded in the Greek psyche. The students and school children who are now laying siege to police stations and trying to bring down the government are undergoing a rite of passage,' the BBC's Malcolm Brabant reported. 'They may be the iPod generation, but they are the inheritors of a tradition that goes back centuries,' he added, before turning to describe the 'current wave of violence' testing the limits of social stability. Brabant, like other journalists on the scene, was struggling to make sense of events which defied easy explanation. Most news accounts agreed that the spark that ignited the student protest was the shooting of a 15-year-old student, Alexandros Grigoropoulos, by a police officer – for no apparent reason – on the evening of 6 December, 2008 in the centre of Athens. The incident received extensive coverage, and 'near-universal condemnation' (Gemenis, 2008), in the Greek media. Where accounts differed was at the level of context; that is, when presenting the details necessary to enable distant audiences to understand the nature of the ensuing crisis, as well as its larger significance. Brabant's references to the 'Greek psyche' as having a predisposition to rebellion were one way to explain the forces involved; another was to frame events historically, where previous examples of police brutality were held to be suggestive of a larger pattern over recent decades. Still another strategy, however, revolved around the gathering of insights being generated via social networks amongst the protestors themselves.

Pavlos Tsimas (2008), a commentator with Mega TV in Athens, described his sense of how events were unfolding in a speech to the Global Forum for Media Development shortly thereafter:

> [At] 9:00 in the evening of Saturday a boy was shot dead. For no reason. In cold blood.
>
> I learned about the fact 80 mintues [sic] later by email. Turned on the TV set and there was nothing on. Just commercials and nice shows. I turned to the Internet and there in some blogs extensive coverage of the event. I kept receiving messages. The clock struck midnight. People took to street to protest the murder. Victim's name nobody knew.
>
> Even radio stations were late to get the news.
>
> Thousands of people in the street protesting murder of a boy whose name they didn't know. Established media have not yet reported the event. TV stations came in a little late. The next day the newspapers did not carry words of the event with the exception some sport papers that carried the story due to late night printing (due to reporting of a football match).
>
> Greece plunged into the deepest crisis in recent memories – people watched fire burning in neighborhoods and saw smashed windows. Radio and TV stations, most of them choose to open the airwaves non stop with call-in shows where listeners expressed themselves and newspapers tried to find out what else to do. (Tsimas, 2008)

Internet and social networking sites, the latter including Facebook, MySpace and Twitter, proved to be playing a pivotal role in mobilising protestor collaboration in real-time. Bloggers posting accounts of what had happened – together with links to a YouTube video which appeared to contradict police claims – led the way. Photos uploaded to Flickr, via mobile telephones and laptop computers, recorded the violent turmoil. Twitter – via tags such as #griots – relayed reports of what was happening on the streets, as well as information for protestors to help co-ordinate their efforts – that is, to relay details about where to meet, what to do, and how best to protect themselves. 'Several Greek Web sites offered protesters real-time information on clash sites, where demonstrations were heading and how riot police were deployed around the city,' Paul Haven (2008) reported for AP. 'Protest marches were arranged and announced on the sites and via text message on cell phones.' Here it quickly became apparent not only that the shooting needed to be situated in relation to protracted civil unrest in Greek public life, but that the protestors themselves did not form a single, monolithic group. The majority of those involved were students, primarily from secondary schools, engaged (peacefully, in the main) in marches and rallies. A second, smaller element, was composed of groups of 'anarchists' (or 'koukoulo-foroi' – 'the hooded ones') intent on seizing the moment to articulate dissent in any way possible – which included torching cars and smashing shop windows. Not surprisingly, the actions of the latter figured prominently in news reports. 'Without a doubt,' Alexis S (2008), a 'peaceful protestor from Thessaloniki' observed, 'such coverage focused on the most sensational, frightening, collective and dramatic cases of the countless incidents that took place in Greece's major cities since Saturday night.'

For journalists looking beyond sweeping claims about 'the iPod generation' – or the '€700 generation' (a phrase used by some to describe the modest monthly wage they hope to earn after university, if they gain employment) – it was important to connect with young people to better understand what they were feeling, and why. Traditional news media, Andrew Lam (2008) noted, 'were trying to play catch up in a world full of Twitterers and bloggers,' a challenge made worse by the need to 'filter real news from pseudo news' under intense time pressure. Social network sites proved to be especially valuable resources in this regard. In addition to providing live reports, personal accounts, photos and videos (most in English, if not in Greek), they afforded journalists with a different level of connection. This level was described by *The Economist* as being indicative of a 'new era of networked protest':

> A tribute to the slain teenager—a clip of photos with music from a popular rock band—appeared on YouTube, the video-sharing site, shortly after his death; more than 160,000 people have seen it. A similar tribute group on Facebook has attracted more

than 130,000 members, generating thousands of messages and offering links to more than 1,900 related items: images of the protests, cartoons and leaflets.

A memorial was erected in Second Life, a popular virtual environment, giving its users a glimpse of real-life material from the riots. Many other online techniques—such as maps detailing police deployments and routes of the demonstrations—came of age in Athens. And as thousands of photos and videos hit non-Greek blogs and forums, small protests were triggered in many European cities, including Istanbul [and] Madrid. Some 32 people were arrested in Copenhagen (*The Economist*, 18 December 2008).

For Evgeny Morozov (2008), the networked protest in Greece provided a 'glimpse of what the transnational networked public sphere might look like,' and as such signalled the rise of a new global phenomenon. Readily acknowledging that the internet has helped to make protest actions more effective in the past, he nonetheless believed that what happened in this case was 'probably the first time that an issue of mostly local importance has triggered solidarity protests across the whole continent, some of them led by the Greek diaspora, but many of them led by disaffected youth who were sympathetic of the movement's causes.'

While their actions were widely condemned by the authorities, the success of their intervention – its ethos spelled out in banners such as 'Stop watching, get out into the streets' – was such that then Conservative Prime Minister Costas Karamanlis conceded in a speech to parliamentary colleagues: 'Long-unresolved problems, such as the lack of meritocracy, corruption in everyday life and a sense of social injustice disappoint young people' (cited in *BBC News Online*, 17 December 2008). Few had anticipated the scope and intensity of the student outrage, nor their conversancy in new forms of digital communication. Journalists were as surprised as anyone else, a point underlined in a headline from the Reuters (2008) news agency: 'Protestors rule the web in internet backwater Greece.' More to the point, the crisis provided evidence that every citizen could be a front-line correspondent, a prospect which called into question the viability of the mainstream news media – not least their capacity to set the news agenda (see also Matheson and Allan, 2010). New media – ranging from established blogs and forums to 'radicalised' Facebook pages – 'directed the flow of information' in the larger 'struggle for meaning,' Maria Komninos and Vassilis Vamvakas (2009) argued, leaving the major news organisations, especially the commercial television channels, 'obliged to report on and to follow, basically, the information originating in the internet.'

Further instances of young people's use of social media for organisational and news-making purposes – sometimes dubbed a 'Twitter protest' or 'Twitter revolution' – attracted periodic attention in the mainstream media over the following months. In Moldova in April 2009, for example, young protestors took to the streets to condemn what they regarded as election fraud involving members of the Communist Party. 'We decided to organise a flash mob for the same day using Twitter, as well as networking sites and SMS,' explained one of the organisers; 'we expected at the most a couple of hundred friends, friends of friends, and colleagues,' she added. 'When we went to the square, there were 20,000 people waiting there. It was unbelievable' (cited in Stack, 2009). Violent clashes ensued, with tweets relaying images and and video clips of confrontations with police, as well as the destruction of windows and furniture in the Parliament building. Shortly thereafter, then President Vladimir Voronin relented, agreeing to a recount of the votes. Social networking strategies proved similarly indispensible during the G20 summit in London the same month. Amongst the estimated 35,000 people demonstrating peacefully were a small number of protestors, some involved with anarchist groups, intent on violent confrontations with the police. 'Amateur' still and video imagery, much of it shot using mobile telephones, recorded several clashes in shocking detail. One incident, in particular, ignited a major controversy, namely the actions of a police officer knocking passerby Ian Tomlinson to the ground. Tomlinson, a newspaper seller, collapsed and died after being hit by a baton. The Metropolitan police's initial denial that an officer had been involved was flatly contracted by evidence revealed by *The Guardian* six days later, namely a video clip documenting the assault handed over to it by an American visitor to the city. For journalist Nik Gowing (2009), the citizen bystander who 'happened to bear witness electronically' represented a telling example of how non-professional 'information doers' were 'driving a wave of democratisation and accountability' redefining the nature of power. 'The new ubiquitous transparency they create,' he observed, 'sheds light where it is often assumed officially there will be darkness.'

June of that year saw major public demonstrations in the aftermath of Iran's disputed presidential election, with many young Iranians performing roles akin to citizen journalists in order to document what was happening (Western journalists having been barred from reporting the protests). Amongst the images assuming an almost iconic status were those taken from grisly mobile telephone footage of 27-year old Neda Agha Soltan bleeding to death on the street (she had been shot in the chest, reportedly by a Basij paramilitary). Relayed to the world's news media, this 'amateur' footage captured by a bystander and uploaded to the web transformed Neda into a symbol of the opposition, galvanising support in Iranian diasporas as well as focusing international attention. In July, demonstrations by Uighur protestors (a Turkic-speaking Muslim group) in the western region of Xinjiang, China were met by police officers wielding fire hoses and batons, sparking 'ethnic riots' that reportedly left 156 people dead and more than 800 injured. The central government, in the words of a *New York Times* reporter, took 'the usual steps to enshrine its version of events as received wisdom: it crippled Internet service, blocked Twitter's micro-blogs, purged search engines of unapproved references to the violence, saturated the Chinese media with the state-sanctioned story' (Wines, 2009). Nevertheless, once again young people's use of social media succeeded in offering real-time firsthand reportage, even though much of it was removed by official censors shortly after it was posted.

These are a small number of examples drawn from a myriad of alternatives leading up to the events currently unfolding in the 'Arab Spring' discussed at the outset of this chapter. While much of the news coverage of these protests has highlighted how young people have learned to exploit the capacity of social networking tools to advance their causes, pressing questions emerge about the nature of civic agency being engendered.

Conclusion: New Agendas

In seeking to explore 'global news literacy' in a digital age, familiar assumptions regarding citizenship need to be examined afresh. Peter Dahlgren's (2009) reformulation of 'civic cultures' is helpful in this regard, I would suggest, because of the way it attends to the conditions giving shape to civic agency as a dynamic process of identity formation. 'Civic cultures,' he writes, 'refer to cultural patterns in which identities of citizenship, and the foundations for civic agency, are embedded' (103). To the extent they generate a compelling sense of 'we-ness,' it follows, they 'operate at the level of citizens' taken-for-granted horizons in everyday reality,' which necessarily entails examining 'those features of the socio-cultural world that serve as preconditions for people's actual participation in the public sphere and political society' (Dahlgren, 2009, 104-105). While an array of factors impact upon civic cultures, Dahlgren suggests that 'family and schools lay a sort of foundation,' before pointing to the influence of 'group settings, social relations of power, economics, the legal system, and organizational possibilities,' amongst others. Singled out for attention in this regard is the 'ever-evolving media matrix,' which makes possible new kinds of civic practices while, at the same time, demanding new skills for citizenship. Interactive media, in particular, offer significant resources for civic identities, especially at the level of lived experience where the 'dialectical interplay of possibilities and their actualizations' construct new contexts of use. 'The ease of interacting, reformatting, remixing, adding on to existing texts, and so forth,' he contends, 'promotes the participatory uses of these technologies – and alters forever the traditional premises whereby mass audiences receive ostensibly authoritative, centralized information in a one-way manner' (Dahlgren, 2009, 154). Nevertheless, Dahlgren cautions, it is simply too early to say how the social transformations under way will be incorporated into civic cultures. In calling for a 'realistic grasp' of the complexities involved, he acknowledges that there is little prospect that digital technologies will deliver a 'quick fix' or 'shortcut to democracy' anytime soon.

Just as the availability of news and information does not in itself ensure an informed citizenry, it is similarly apparent that no corresponding relationship can be presumed to exist between young people's involvement in social networking and their aptitude for civic participation. One in no way prefigures the other, but this is not to deny that individuals conversant in the uses of technologies widely associated with 'Web 2.0' will be well-placed to advance personalised, affective forms of engagement with issues they consider relevant to their concerns (see also Allan and Thorsen, 2009; Moeller, 2009). Taken together, the types of political intervention highlighted over the course of this chapter may be read as broadly indicative of an emergent, uneven and frequently contested ethos of digital citizenship. This is not to suggest that those involved self-identify with specific roles, duties or obligations consistent with traditional (that is, prescriptive) ideals of democratic responsibility. Discourses of citizenship may or may not resonate with these young people's performative identities, let alone their sense of belonging within a shared community. Instead, their actions are more likely to be defined in these terms by commentators anxious to reaffirm 'real' tenets of political mobilisation and protest in the face of 'virtual' alternatives. Consequently, I would argue that citizenship must be rethought in a manner alert to its multiple, socially contingent re-inflections within a new media ecology where such dichotomies have long ceased to claim a conceptual purchase.

Current discussions about how media education can most effectively address the changing imperatives of this evolving mediascape have much to gain from revisiting earlier instances where rapid technological change was criticised for ushering in undesirable forms of media content. In the case of Leavis and Thompson's (1933) intervention, echoes of their call to resist the lure of shallow gratifications can be heard to reverberate in reservations expressed about young people's media preferences, even where care is taken to avoid overtly elitist claims about harmful influences. The disdain frequently shown for students' use of Twitter or Facebook, for example, can take the form of a 'cyber-scepticism' that chastises them for being isolated from reality – in effect, a generation of 'slacktivists' too lazy to engage in face-to-face communication, let alone inform themselves about political issues with a view to getting involved (see also Gerodimos and Ward, 2007; Morozov, 2011; Turkle, 2011). Related criticisms about emotional dislocation, detachment, inauthenticity, and the like, would not sound out of place in *Culture and Environment* published all those years ago. At risk of being overlooked, however, is the extent to which social networking is intimately interwoven into the fabric of young people's everyday lives, as well as the reasons why connectivity is so deeply valued. Dahlgren's (2009) conception of 'civic cultures' reminds us that there are many ways of being a citizen, of 'doing democracy' – civic identities, in his words, 'are not static, but protean and multivalent' (119). Media education, it seems to me, ignores these lived contingencies at its peril.

Efforts to rethink civic engagement, I would suggest, need to better understood how personal experience gives shape to the ways young people relate to their communities beyond 'citizenship' narrowly defined. It is in the gaps, silences and fissures of more traditional definitions that the basis may be discerned for envisaging alternative networks of civic participation firmly situated in the politics of the everyday. Important here is the need to discern the basis by which young people may be encouraged to embrace their civic selves, that is, to recognise themselves as prospective participants contributing to democratic cultures in self-reflexively meaningful, purposeful ways. In opening up a wider debate concerning how best to improve the quality of news literacy in this regard, it follows, every effort must be made to ensure young people's views, experiences and perspectives inform the ensuing dialogue about what it means to be a citizen in a digital age.

Acknowledgement

For offering helpful comments on a first draft, my thanks to Paul Mihailidis as well as Cynthia Carter, Chindu Sreedharan and Einar Thorsen.

REFERENCES

Allan, S. (2002). "Media Studies and the Knowledge Society," *Southern Review: Communication, Politics and Culture*, 35(2): 100-112.

Allan, S. and Thorsen, E. (eds.). (2009). *Citizen Journalism: Global Perspectives*. New York: Peter Lang.

Brabant, M. (2008). "Rebellion Deeply Embedded in Greece," BBC News Online, 9 December.

Buckingham, D. (2003). *Media Education: Literacy, Learning and Contemporary Culture*. Cambridge, UK: Polity Press.

Carter, C. and Allan, S. (2005). "Hearing Their Voices: Young People, Citizenship and Online News," in A. Williams and C. Thurlow (eds.), *Talking Adolescence: Perspectives on Communication in the Teenage Years*. New York: Peter Lang, 73-90.

Dahlgren, P. (ed.). (2007). *Young Citizens and New Media*. London: Routledge.

——. (2009). *Media and Political Engagement*. Cambridge, UK: Cambridge University Press.

Economist, The. (2008). "Rioters of the World Unite," *The Economist*, 18 December.

Gemenis, R. (2008). "Greece in Turmoil: Riots and Politics," *OpenDemocracy*, 10 December.

Gerodimos, R., and Ward, J. (2007). "Rethinking Online Youth Civic Engagement: Reflections on Web Content Analysis," in B.D. Loader (ed.), *Young Citizens in the Digital Age*. London: Routledge, 114-126.

Gowing, N. (2009). "Real-Time Media is Changing Our World," *The Guardian*, 11 May.

Guedes Bailey, O., Cammaerts, B., and Carpentier, N. (2008). *Understanding Alternative Media*. Maidenhead, UK: Open University Press.

Hassan, R. (1999). "Globalization: Information Technology and Culture in the Space Economy of Late Capitalism," *Information, Communication and Society* 2 (3): 300-317.

Haven, P. (2008). "Greek-Inspired Protests Spread across Europe," Associated Press Worldstream, 12 December.

Hobbs, R. (2010). "Digital and Media Literacy: A Plan of Action," a White Paper on the Digital and Media Literacy Recommendations of the Knight Commission on the Information Needs of Communities in a Democracy, Washington DC: The Aspen Institute.

Ito, M. (2010). "Lessons for the Future from the First Post-Pokémon Generation," *Nieman Reports*, 64(2): 18-20.

Kellner, D. (2002). "New Media and New Literacies: Reconstructing Education for the New Millennium," in L.A. Lievrouw and S. Livingstone (eds.), *Handbook of New Media*. London: Sage, 90-104.

Komninos, M. and Vamvakas, V. (2009). "The Role of New Media in the December 2008 Revolt in Greece", *Encyclopedia of Social Movement Media*, J. Downing (ed). London: Sage.

Lam, A. (2008). "Letter from Athens," New America Media, 16 December.

Leavis, F.R. and Thompson, D. (1933). *Culture and Environment*, London: Chatto and Windus.

Lewis, J., Inthorn, S., and Wahl-Jorgensen, K. (2005). *Citizens or Consumers? What the Media Tell us about Political Participation*. Maidenhead, UK: Open University Press.

Lievrouw, L.A. and Livingstone, S. (eds.). (2002). *Handbook of New Media*, London: Sage.

Lister, M., Dovey, J., Giddings, S., Grant, I. and Kelly, K. (2008). *New Media: A Critical Introduction*, Second edition, London: Routledge.

Loader, B.D. (ed.). (2007). *Young Citizens in the Digital Age*. London: Routledge.

Matheson, D. and Allan, S. (2010). 'Social Networks and the Reporting of Conflict,' in R. L. Keeble, J. Tulloch and F. Zollmann (eds.), *Peace Journalism, War and Conflict Resolution*. New York: Peter Lang, 173-192.

Messenger-Davies, M. (2010). *Children, Media and Culture*. Maidenhead, UK: Open University Press.

Mihailidis, P. (2009). *Media Literacy: Empowering. Youth Worldwide*. Washington, DC: Center for International Media Assistance.

Mitchell, A. (2010). 'Revealing the Digital News Experience: For Young and Old,' *Nieman Reports*, 64(2): 27-29.

Moeller, S. (2009). *Media Literacy: Citizen Journalists*. Washington, DC: Center for International Media Assistance.

Morozov, E. (2008). 'The Alternative's Alternative,' *OpenDemocracy*, 29 December.

——. (2011). *The Net Delusion: How Not to Liberate the World*. London: Allen Lane.

Pew Research Center's Internet & American Life Project. (2011). 'Social Networking Sites and Our Lives.' Washington, DC. http://pewinternet.org/Reports/2011/Technology-and-social-networks.aspx

Reuters. (2008). "Protestors Rule the Web in Internet Backwater Greece," *Reuters Global News Journal*, 18 December.

Rosen, L. (2010). "Understanding the iGeneration: Before the Next Mini-Generation Arrives," *Nieman Reports*, 64(2): 24-26.

S, A. (2008). "Eye Witness from Thessaloniki," *openDemocracy*, 12 December.

Sefton-Green, J. (eds.). (1998). *Digital Diversions: Youth Culture in the Age of Multimedia*. London: UCL Press.

Stack, G. (2009). "Twitter Revolution" Moldovan Activist Goes into Hiding," *The Guardian*, 15 April.

Tsimas, P. (2008). Transcript of speech to Global Forum for Media Development, posted by Andrew Lam on 'Chez Andrew,' *New America Media*, 10 December.

Turkle, S. (2011). *Alone Together: Why we Expect More from Technology and Less From Each Other*. New York: Basic Books.

Warnick, B. (2002). *Critical Literacy in a Digital Era*. Mahwah, NJ: Lawrence Erlbaum.

Wines, M. (2009). 'In Latest Upheaval, China Applies New Strategies to Control Flow of Information.' *The New York Times*, 6 July.

Chapter 2 -
Media Literate "Prodiences":
Binding the Knot of News Content and
Production for an Open Society

MANUEL ALEJANDRO GUERRERO & MÓNICA LUENGAS RESTREPO
Universidad Iberoamericana, Mexico City, Mexico

Introduction

Three crucial changes in recent decades—an unprecedented technological revolution, the fall of Communism, and strong deregulation trends worldwide—have transformed the ways in which news and information are produced, disseminated and received. One outcome of such changes is the incidence that individuals now have in these processes. *Audiences* have transformed into what this chapter develops as *prodiences*: active citizens intervening in the shaping of content through the use of digital media technologies and platforms.

As a result, a steady growth in literature dealing with new models for journalism has emerged, discussing what various factors are actively re-shaping news and information in the 21st century. However, most fails to link these factors to two crucial aspects of present day news constructs: a wider theoretical debate on the roles of news and information for democracy in digital spaces, and media literacy approaches to civic participation, production, and activism.

The aim of this chapter is to present new models for content production both by anchoring it to liberal theoretical principles addressing the function of information for civic society and by considering media literacy as an integral part of these models in relation to the new roles played by *prosumers*[1] in the production of information. The new models proposed herein reconsider both a normative approach for news in open societies and the factors that may ac-

---

1 According to Ritzer and Jurgenson (2008), 'prosumption' involves both 'consumption' and 'production' whose fusion is favored by a series of social changes, especially—though not exclusively—those related to the Internet and Web 2.0.

tually shape content production. This discussion, then, is about information content and production, two constructs that are strongly related but differ in key areas of today's digital information landscape. Media literate prodiences, this chapter contends, may be their link.

This chapter will develop and explore the idea of media literate prodiences in four parts. In Part One we propose an operationable version of the classic liberal ideal-type role that news should play in an open society. In the Part Two, we use this ideal-type for presenting a model linking factors that shape content production with news content itself (*production factors-information content*). In Part Three we assess how unprecedented technological revolutions, the fall of Communism and the strong deregulation trends have all impacted traditional factors that shape content production and generate the conditions for the emergence of today's prosumers. Finally, in Part Four, we discuss how media literacy can incorporate the role of the prosumer into a theoretical *production factors-information content* model.

Rethinking the Liberal Normative Approach of News/Information for Open Societies[2]

On January 22, 2010, journalist Dave Iverson hosted a debate on NPR titled "The Death and Life of American Journalism", in which the challenges to 21st century journalism were discussed with John Nichols and Robert McChesney, whose latest book gave the title to the show. During the program both authors agreed that journalism is not only related to democracy, but fundamental to it, since, as Nichols stressed, it is the "duty of newspapers and other news outlets to provide their readers and viewers with essential and well-reported information" (Benfield, 2010). In fact, in modern Western countries the relationship between journalism and democracy is believed to be rooted in the defense of liberal views on freedom of expression and freedom of the press, which can be traced back to Milton's *Areopagitica* (1644).[3]

---

2 The term is, of course, referred to the work of Karl Popper (and his references to Bertrand Russell) in which liberty and freedom, tolerance and human rights are observed in a society's public life and whose government is replaced by institutional mechanisms that reject the use violence (Popper 1963 [1945]). It could be argued that today, most of an open society's premises may be better fulfilled in modern liberal democracies.

3 The complete title is quite eloquent for our matters, Areopagitica: A speech of Mr. John Milton for the liberty of unlicensed printing to the Parliament of England. Later, liberal thinkers, like John Stuart Mill –especially in his *On Liberty* (1859)– defend the idea that freedom of the press is necessary because it enables a true marketplace of ideas, it permits open debates that

In an already classic work, Alexander Meiklejohn argues that since democracy implies popular sovereignty, its citizenry—responsible for rotating governments through elections—requires an adequate amount of information in order to make informed choices. According to Meiklejohn, freedom of expression and freedom of the press serve two functions in a democracy: an *informative function* that allows for a necessary information flux; and a *critical function* that guards against the abuse of power and contributes to the critical evaluation of government (Meiklejohn, 1960). From this standpoint, it is possible to assume then that the ultimate goal of news content in the public sphere is to strengthen democracy by providing information to citizens, serving as a watchdog of those in power, and reflecting an open arena of free debate. This view constitutes the basis of the normative liberal conception of the roles of journalism, and of the informational media in modern democracy (Siebert, Peterson & Schram, 1956; Michnik & Rosen, 1997; Ungar, 1990).

It is true that such a conception of democracy and information is overly idealized and subject to strong criticism (see, for instance, Curran, 2002; Curran & Seaton, 2003; Dahlgren & Sparks, 1997). Notwithstanding, those supposed roles of journalism—and of the media as "information providers"— serve to distinguish an *ideal type* (in Weberian terms) of information in a democracy from what might be expected in other kinds of political regimes. Moreover, this ideal type could help contrast even what occurs within democracies themselves in terms of the watchdog role of journalism, the need for open and balanced civic debates, or the quality of information conveyed by mainstream media to citizens. This liberal ideal type is still relevant as a framework to understand rights, freedoms and claims for better quality information in a pluralistic society.

However, to assume that the ultimate goal of news in the public sphere is to strengthen democracy, though a noble goal it may be, requires an explanation about how it may actually work. If, from a liberal standpoint, we maintain the journalistic role of a watchdog, we ultimately assume journalism as enhancing accountability in public life. Accountability is a term that strongly relates to responsibility, liability and answerability, and implies the obligation to report, explain or justify something; and further to be responsible for something before someone else (RHDEL, 1987). Schedler describes accountability

---

lead to truth, it prevents both the tyranny of a majority and the corruption and abuse of governments (Mill, 1989; also Alexis de Tocqueville in his *Democracy in America* presents similar arguments in favor of a free press).

as a situation in which "A is accountable to B when A is obliged to inform B about A's (past or future) actions and decisions, to justify them, and to suffer punishment in the case of eventual misconduct" (Schedler, 1999). Jabbra & Dwivedi (1988) identify different types of accountability, including moral, administrative, managerial, market, political, legal, constituency relation and professional.

For the purposes of this work, we employ four main types of accountability that are expected from news in an open society: market, moral, administrative, and political. While the boundaries for these categories are not always clearly defined, exploring each separately here will help build a more nuanced understanding of accountability in the context of this chapter:

-   Market accountability entails the coverage of stories related to the defense of consumers, the promotion of better competition practices, and the disclosure of corporate misbehavior, including stories that may involve conflicts of interest between the economic agendas of media organizations and professional journalistic practices.
-   Moral accountability focuses on human interest and on information that is presented to show, defend and/or promote cases that may endanger human life, dignity, freedoms, and basic individual and collective rights.
-   Administrative accountability implies covering stories related to the ways in which "a public agency or a public official fulfills its duties and obligations, and the process by which that agency or public official is required to account for such actions" (Jabbra & Dwivedi, 1988, p. 5). In simple terms, all news that conveys information to monitor an administrations actions.
-   Political accountability can be said to provide information for citizens to have the possibility to sanction, usually, though not exclusively, through electoral procedures, the general performance of governments and other elected officials, as well as to assess their answerability for their actions and decisions (Bellamy & Palumbo, 2010; Schedler, Diamond & Plattner, 1999; Przeworski, Strokes & Manin, 1999).

In this way, the liberal ideal-type proposed through these four accountability types may serve not only to distinguish the limitations of news content in non-democratic systems, but also to assess the challenges that news faces in

actual democratic and transitional polities. In this way, a richer liberal normative "ideal type" may be useful to assess a reality full of tones of gray, both in
non-democracies and democracies. It is true that, as an ideal type, news content is expected to meet all four kinds of accountability in full-fledged democracies. However, degrees of accountability can be attained by the news even in
countries and contexts that are not entirely democratic. This makes the ideal
type to become operational. For instance, in Poland during Communism,
many journalists used each of the crises of the Communist Party rule (1956,
1968, 1970, 1976, 1980-1) to imply that more open mass media could help
alleviate the tensions before they led to societal unrest (Curry, 1989). Many of
them were not advocating the end of Communism, but wanted to transform
some spaces within it and obtain more independence and autonomy in their
profession. Thus, since the times of Gomulka:

> The new party leadership [...] knew that it needed the support of the print journalists
> to maintain its own position and to prevent a Soviet invasion. This informal bargain
> and the knowledge that they also had the support of the populace gave the Polish
> journalists, already equipped with an established professional ethic, some power in
> the defense of their work. The journalists repeatedly tested the boundaries of the
> permissible, an arrangement that was fostered by the censorship process. The result
> was a system that allowed for more open reporting and discussion of issues, which
> over time inevitably raised questions about the authority and knowledge of the party.
> The press was therefore part of the system, and yet apart from it: The party needed
> the press. (Johnson, 1998,109-110, 112)

Another example comes from private television in Brazil during the military dictatorship (1964-1985). Though the O Globo multimedia empire was
created during this period (and in many ways O Globo top managers and owners supported that regime), by the early 1980s the corporation decided to shift
its allegiance in favor of political openness through direct elections—the direitas
já[4]—in a moment when the military were still in control of power (Singer,
1994).[5] Perhaps even China's press could show some examples of administra-

---

4 Direitas ja was the campaign in favor of holding direct popular elections for the election of
Joao Baptista Figueiredo to the presidency of Brazil. It is important to note that, though
O'Globo supported the military, since the times of Ernesto Geisel it decided to establish an
alliance with the military soft-liners of the regime.
5 In Brazil, it is also remarkable the almost spontaneous way in which O'Globo directors decided to shift their preferences. According to two consistent observers of Brazilian media, Guimaraes and Amaral shifted from practically ignoring the popular protests to becoming strong
supporters of the electoral process, encouraging the people to join (Amaral and Guimaraes,
1988).

tive accountability when they cover cases related to corruption in the public administration, or to bureaucratic inefficiencies, or when journalists resort to criticisms regarding regional governmental reactions during natural disasters.[6] While news may ideally meet fully all four kinds of accountability in functioning democracies, this does not exclude the possibility that in other kinds of polities information could attain some level of accountability.

In brief, we propose an operationable version of the classic liberal ideal-type of the roles that news content should play in the public sphere. To fully develop this model we must discuss the factors that may help shape the production of information, taken either as a "natural product" of journalism or of the media's role as information providers. In the following section we link the liberal ideal-type proposed here with the factors that may actually shape information production into a more comprehensive model of information processing.

Modeling News Production Factors: The 20th Century Model

Perhaps one of the best known systematic efforts to revise the forces that work to shape media production is Shoemaker and Reese's work *Mediating the Message* (1996) in which, through discussing different classical contributions to the field—from White (1950), Tuchman (1978), Gitlin (1979) to Gans (1980), among others—classify at different levels of analysis the diverse factors that shape the production of media content into an organized model called the "hierarchy of influence on media content" model. These factors work at:

-   *Individual level*: in which content selection is influenced by individual media workers and communicators in certain key positions (i.e., "gatekeepers").
-   *Routines level*: in which the selection of content is marked by the nature of editorial practices and by the forms of organizing the work in the media outlet.
-   *Organizational level*: in which content is affected by group decision-making.

---

6 Apparently, administrative corruption has been a widely covered topic by the Chinese press, at least since the times of Deng Xiaoping, when it was created in 1978 the Central Discipline Inspecting Committee under the Central Committee of the Chinese People's Party specifically to deal with corruption and misconduct of members of the CPP (Harris, 1988; Datta-Ray, 1998; Chen, 2000).

- *Extramedia level*: in which content is influenced by social institutions, markets, interest groups and audiences.
- *Ideological level*: in which content is defined by ideological positions.

The model (Figure 2.1) is represented as different concentric circles in which the core is the "individual level" and the outmost circle is the "ideological level", which in a way comprises or creates the broadest context for the rest. In a more recent work, Shoemaker and Vos (2009) basically outline the same five levels of influence on the production of media content, though they change the name of the last two: instead of calling the fourth "extramedia", they call it "social institution", and instead of "ideological", they name it "social system" (33-34). Now if we superimpose on these "hierarchy of influence" models our previous liberal ideal-type of news contents, the outcome can be represented by conditions that show what could be normatively expected by *news content* in terms of accountability at each of the different levels that shape *information production* (Figure 2.1).

For the sake of simplicity we may gather into a single category the first three levels of Shoemaker and Reese's model and call it "corporate level" as it implies individual media workers' profiles, criteria of professionalism, practices and routines, and also the organizational structures and divisions of labor within. Normatively, we might expect at this level that media workers and communicators meet professional and ethical standards,[7] that there are clearly established norms and routines that define and divide functions, that reporters and journalists have an accepted degree of autonomy from editors and owners and, in turn, editors enjoy autonomy from owners. Now, at the extramedia level, where content is influenced by social institutions, markets, interest groups and audiences, normatively we might expect that market rules are clear, competition is fostered, monopolies/oligopolies are banned, lobbying rules and other forms of interest group pressing are also clearly limited, and the audiences/readers have some procedures to make their voices heard. Finally, at the ideological (or social system) level, where content is shaped by ideological principles, normatively we might expect that liberal democracy is reflected in laws, regulations and other principles that both: grant independence of the media system from the political system, like establishing clear and open procedures for awarding licenses, or using gov-

---

7 In terms of education, formation, ethical standards, detachment, unbiased attitudes in coverage, reporting or presenting information, and so forth.

ernment funds for support or publicity, and; defend freedom of speech, of information, and of the press, and protect the basic principles of professional journalistic practices (right of consciousness, right not to disclose sources, and the like).

If these ideal-type conditions apply, it is then possible to attain accountability through news because journalists and reporters practicing their profession within organizations that grant them autonomy and clear cut labor standards could have the proper conditions to cover stories of human interest and to watch over the performance of the public administration's policies, practices and officials without the interference of other interests or pressures within the media corporation. Also, professional journalists could work within the proper conditions to cover corporate misbehavior, funding practices, abuses against consumers (audiences and readers) and so on. Finally, political accountability can also be met through proper legal conditions for journalists to monitor and provide citizens with prompt, useful and valid information. Thus, three broad levels of factors that shape media production may favor accountability through the coverage of information.

It is naïve to think of a clear-cut scenario in which one specific set of production factors fosters only one specific type of content, however based on Shoemaker & Reese's (1996) concentric circles, we may say that broadly when we have professional journalists working with clear routines—i.e., at the "corporate level"—they will try to cover all sorts of stories of human, economic, social and public interests. Nevertheless, they will cover economic and market stories successfully only if they, as well as the editors and the organization in which they work, can be autonomous from corporate and interest group pressures, i.e. at the "extramedia level". In the same way, they will succeed in covering administrative and political stories only if proper rules and protections grant them their basic rights and freedoms before the political system (government, parties, specific politicians, public agencies, etc.), i.e. at the "ideological level". In this way, if it is true that it is not realistic to determine a one-to-one relation between these factors, broadly speaking each of the levels that shape information production may better promote a certain kind of accountability in news content, as it is represented in Figure 2.1.

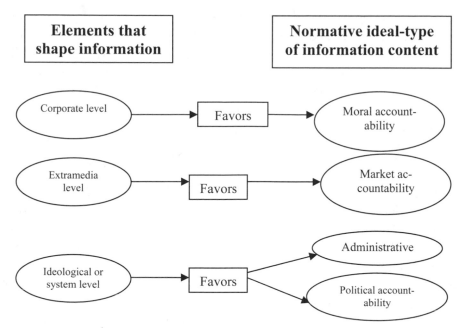

Figure 2.1 - 20th Century Model of Relation "Production Factors-Information Content" [Source: Schumaker & Reese, 1996]

This model could be applicable to most 20th century *production factors-information content* relations. As can be seen in this model, audiences, readers and information consumers are excluded from the (direct) process of generating information. They are end receivers in the process of content production (Figure 2.2).

Figure 2.2 - Roles of Audiences in the Traditional Content Production Process (Courtesy of Authors)

For Shoemaker & Reese's (1996) "extramedia level" and for Shoemaker & Vos's (2009) "social institution level," individuals—acting organized in pressure or interest groups, or as advertisers or audiences—are factors influencing media content production, however they do so indirectly. As interest groups they do so by pressing media organizations in favor of certain interests, as advertisers they

do so by lobbying or collaborating with top managers in favor of certain products, makes or brands, and as audiences they do so when producers realize that
certain information does not achieve expected ratings. Individuals, as media
consumers outside media organizations, have not had the chance to interfere in
content production *directly*. However, the situation has changed in our day and
that is why we now must include individuals in the model of factors that shape
media production today. The question is how? Let us first discuss why.

Structural Changes and Their Impact on Information Production

Over the last three decades significant structural changes have altered the conditions under which information is produced. These structural changes started
in the late 1970s and continued throughout the 1980s. They are mainly cate-
gorized as: deregulation/liberalization trends in the political economy; technological revolutions; and the downfall of Communism with the predominance
of today's consumption-based single ideology of free market. Now, we summarize some of the impacts of these changes into the levels—using Shoemaker &
Reese's terms (1996)— that used to shape information production in the days
before the individual became a content producer.

At the extramedia level, the factors that influenced content production have
been impacted by those three changes in many significant ways. To start with, media
markets are becoming increasingly concentrated as part of global economic trends,
and new gigantic multimedia conglomerates are defining mainstream media markets (Bagdikian 2004). De-regulated markets typically see audiences as consumers
and the pre-eminence of information lies more and more in its capacity to sell and
generate benefits, strengthening the trends to reduce those spaces that cannot translate into profits in the market. As for the possible impact of these trends on information, such profit seeking attitudes may influence the format and content of news
programs, making business criteria the new editorial policy in the newsrooms and
generating different forms of "corporate censorship" (Keane, 1991).

Market-oriented conceptions of the media must take into account the fact
that media are primarily businesses competing in a market fighting for survival
and expansion. For the critics, such features tend to work precisely against the
liberal precepts above, since the consumer, not the citizen, is the most important target (Tracey, 1988). At the same time, the logic of the market dictates
that, though there may be a large variety of newspapers and broadcast outlets,
their content tends to homogenize in the information the media organization
disseminates. A larger variety of channels, for instance, does not necessarily

entail content diversity. Therefore, "the overall programme fare becomes relatively thinner, more repetitious and more predictable than would otherwise be the case. Inevitably, the ratings dominate. But ratings under-represent the opinions of ethnic and regional minorities, gay, lesbians, greens, elderly citizens, socialists and other minorities" (Keane, 1990, 77). Obviously, this has a negative impact on the capacity of the media to critically cover other interests of the corporation to which they belong, affecting their information policies according to corporate interests (Curran & Seaton, 1988; Murdock, 1990).

Heinonen & Luostarinen (2008) identify a general trend in which media schools and universities are generating work-ready professionals that are able to perform diverse tasks without completely identifying with the traditional values of a journalistic profession. The authors say:

> On the basis of their education, they could work equally well in marketing, public relations, advertising, etc. –and that is what they are actually doing, moving freely from one media branch to another. In the long run, this development will make the profile of "a journalist" much more unclear both in the eyes of the audience and the profession itself. In addition, often –though not exclusively–in relation to new media technology, journalistic work processes have changed so that previously relatively well-established job descriptions have been crumbling....(Heinonen & Luostarinen, 2008, 232)

As for possible impacts on news, scholars already identify possible dangers resulting from the commoditization of news (McChesney, 2008). For instance, the discussion emphasizes that television formats and rhythms avoid an in-depth discussion of issues and instead transforms civic debates into mere entertainment, sacrificing content for attractive formats.[8] Other relevant issues raised by a changing working environment are the questions about the costs of investigative journalism and who pays for such reporting. Around the world, less and less media organizations are now capable of financing their own journalists to travel, investigate, and compose stories that are most often problematic to publish in their entirety. In brief, "doing professionally good work in journalism is becoming increasingly difficult" (Heinonen & Luostarinen, 2008, 233).

Another significant change at the corporate level has to do with the ways traditional journalistic routines have been blurred in cross-media platforms and online newsrooms (Silcock & Keith, 2006; Zavoina & Reichert, 2000). In the creation of online news the chain of command, protocols and publishing standards are not yet

---

8 For a discussion of these positions, see Langer 1998, especially his first chapter, "The lament, critical project and the 'bad' new son televisión".

established or vary from organization to organization (Paterson & Domingo, 2008). In this regard, content has been impacted in contradictory ways.

On the one hand, media corporations are trying to cope with the velocity of change, for instance breaking news can be promptly and timely uploaded, stories can be followed up without regard for space limitations, information from all over the world can be almost immediately printed, etc. On the other hand, the majority of media corporations have yet to develop a truly successful business model from their online news divisions. And thus, online journalism is still fraught with the same problems of financing investigative reports, the cutting of expenses and personnel, and the pressures on professionalism. As Heinonen & Luostarinen (2008) say, journalists in the media are caught in the midst of growing pressures emanating from market forces, technology and the public.

Finally, at the ideological or system level, the predominance of a single market-oriented and consumer-based ideology has justified deregulation and privatization policies around the world, which has created a media landscape characterized by two trends. First, the discourse of social or civic responsibility that underpinned the information role of the media has been losing ground to a blatant commercial orientation. Second, there is now an expansion in the number and types of (multi)media outlets, channels, platforms and other interactive devices. The traditional division of media into print and broadcast in which licenses, permissions and registrations were required for operating is today senseless, since there is a growing proliferation of other types of media platforms in unregulated or legally-gray spaces. Countries with traditionally strong public media services are now surrounded by private services; countries with strong commercial broadcasting services are facing niche competition from small internet stations; and in almost all cases, presses are facing their most trying times in over a century. At the macro level competition today is not between media types, but between multimedia service providers and delivery platforms. The impact of these trends on content can be assessed by their commoditization, as the real value of the diversity of multimedia platforms and services lies in their capacity to distribute and commercialize information. Hence, under this logic, content, from entertainment to information must be before everything else, profitable. Figure 2.3 represents the impact of these changes on the levels of information production.

Under these new circumstances, the normative ideal-type of information content has been devalued in relation to a landscape in which content is strongly correlated with ratings, consumption and profits, and less so with formative content, citizenship and civic engagement. The question then be-

comes how is it possible that under these new conditions at the three production levels, the normative ideal-type of information (based on liberal values)
may still be considered a useful parameter for information?

The answer may be found in a fourth level that must now be considered as a
significant factor in the production and dissemination of media content, and that
has appeared mostly due to the technological advancements and justified by market ideology. This fourth level refers to individuals who, thanks to technology,
deregulated spaces and the preeminence of private initiatives, are stepping up into
the content production process. We call these individuals "prodiences," because
they not only receive information, but collect, report, interact, discuss, redefine,
redesign, create and disseminate information both outside mainstream media
organizations and within the new spaces (i.e., blogs, microblogs and social media
platforms). This means a structural transformation that expands beyond the extramedia level and affects the other levels *directly*.

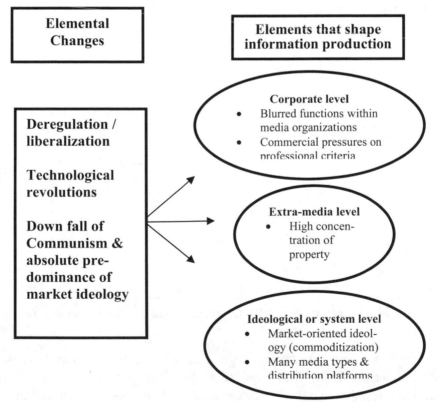

Figure 2.3 - Impact of the Changes on Elements that Shape Information Production (Courtesy of Authors)

Prodiences have a direct incidence at the corporate level, as it is no longer unusual that mainstream media react to and take from the news independent bloggers produce. At one level, prodiences are now able to "produce" content, not at the depth of media organizations, but at an increasingly relevant pace nevertheless. At another level, the practice of citizen or participatory journalism (Lasica, 2003) is questioning professionalism in journalism and other corporate media routines. In many ways the emergence of new (mostly non-institutional) forms of communication are becoming increasingly relevant for news making (Tumber, 2001), and challenging the traditional roles of journalism. Moreover, in some cases individual content producers, especially bloggers, are regarded as trustworthy sources in conflicting situations or when local media organizations are regarded as biased. For instance, the case of Cuban Blogger Yoani Sanchez–named World Press Freedom Hero in 2010–who through her blog "Generation Y" provides glimpses to the otherwise difficult-to-know Cuban daily life conditions.

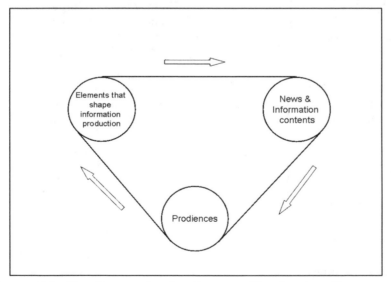

Figure 2.4 - Prodiences in the Content Production Process (Courtesy of Authors)

Prodiences also have an incidence at the ideological or system level as, in many respects, they incarnate a refreshing discourse on freedom of speech and information rights that were once predominantly dictated by mainstream media and governmental agenda-setting. Moreover, whether they are active bloggers or citizens asking questions or commenting via Twitter, new prodiences are providing different and newer views on topics and on the perspectives within which they are discussed. Shayne Bowman and Chris Willis in their work, *We Media* (2003), emphasize that the final aim of participation in the creation of content is the distribution of accurate, independent, reliable and relevant information required by a democracy. Figure 2.4 shows these transformations.

Through this fourth level of prodiences we see a re-emergence of the normative liberal aims of information for democracy. However, this re-emergence is the very reason why we finally need to link this discussion about information production and content with the skills required by these prodiences who are now able to intervene in these processes. This last section argues that media literate prodiences are best suited to attain such goals.

Conclusion: Media Literate Prodiences: Binding the Knot

There are many definitions of media literacy. Most, however, overlap at the individual's ability, on the one hand, to access, analyze, understand and produce contents, and on the other to develop a critical thinking and an awareness in relation to media content (Potter, 2010; Adams & Hamm, 2001; Brunner, 1999; Christ & Potter, 1998; Heins & Cho, 2007; Mihailidis, 2009a; OFCOM, 2005; Rivoltella, 2006; Silverblatt, 2001). Silverblatt insists on this latter point when saying that "media literacy promotes critical thinking skills that enable people to make informed decisions in response to information conveyed through the channels of mass communication" (2001, p. 2). In his *Theory of Media Literacy*, Potter (2004) says that media literacy provides the skills necessary for the individuals to make informed choices about content:

> Media-literate people make their own choices and interpretations. They do this firstly by recognizing a wide range of choices; then, they use their personal, elaborated knowledge structures for context to make decisions among the options and select the option that best meets their own goals. Non-literate people have no choice but to allow the media to make them, because they have few options and the options they do have are given to them by the media. It's rather like getting up in the morning and finding that your significant other has given you two choices of what to wear to the office: a clown suit or a mermaid outfit. It's a choice in the literate sense of the term, but not a real choice. The media give us choices, but not nearly the range of options that we need to have to make the best choice, given our goals. (p. 57)

Critical thinking is key for media literacy since it enables the individuals to ask themselves about what they consume from the different kinds of media contents. Elizabeth Thoman (2010), founder of the *Center for Media Literacy*, explains that media literacy involves three stages of a continuum leading to media empowerment: the first one involving awareness of our own media diet; the second, involves learning skills for critical consumption, specifically related to disclosing and analyzing the frames through which media contents are presented; and the third stage involves asking questions on who produces the information, for what purposes, who profits, who loses and who decides.

Mihailidis (2009b) goes one step forward when he affirms that "across all levels of education, media literacy is increasingly seen as a core component in preparing the future public for active and engaged citizenship". Following Aufderheide (1993), the author stresses that media literacy not only seeks to enhance students' media analysis skills, but also their critical understanding of media's larger political, cultural, and ideological implications. Then, Mihailidis (2009b) proposes a general frame "for exploring how media can bridge cultural, political, and ideological divides" that is called the "Five A's" of media literacy that "enable a continuum starting with an understanding that there is no democratic society without access to information, and concluding with the idea that in today's hypermedia environment, we all have the ability to be active participants in global communities". The Five A's are: **access** to the media technologies as the most fundamental necessity for a global media literacy framework, since "without access to information, a democratic society would cease to exist in its current form"; **awareness** of media power; **assessment** of the ways in which they cover topics and events; **appreciation** of their role in generating civil society; and, **action** to foster better quality communication between cultural, political and social divisions.

As can be seen, these Five A's are crucial for generating media literate individuals, or as we have called them here, prodiences. In this regard we want to emphasize the role of "action", since as a fundamental characteristic of today's active content generators, individuals are perhaps more active than ever in the creation–and re-creation–of content. Following again Mihailidis (2009b) in relation to "action" he says that:

> Never before have there been so many avenues for active participation in global dialogue as there are now. Internet and new media technologies have enabled new means for media production and activism. Global media literacy must teach how such newfound avenues for expression can empower people to take action. How can I use media to have a voice? What are the avenues for active participation in civil society? How much do I participate in the creation of cultural understanding, tolerance, and global progress? How can new media empower active and engaged citizenship? (40)

In this sense, new active individuals as prodiences constantly receive, collect, report, interact, discuss, redefine, redesign, create and disseminate content both outside mainstream media organizations and within the new spaces provided by them. Figure 2.5 represents the way in which media literate prodiences, as key factors within the content production process, may attain the normative liberal aims of information for an open society.

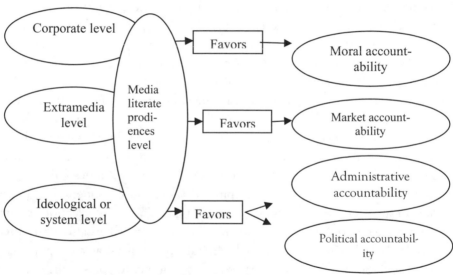

Figure 2.5 - 21ˢᵗ Century Model of Relation "Production Factors- Information Content" [Courtesy of Authors]

The technological revolutions, deregulatory policies and spaces fostering private initiatives over public policies have generated the conditions that favor individuals to intervene in direct ways in the creation of information, and have at the same time challenged the logic of the traditional factors that used to shape content. That is why it is important to stress the relevance that individuals acquire today in reconfiguring the new debates in the public sphere.

On the one hand, as we discussed before commercial pressures and deregulatory tendencies forced media organizations to accommodate new more fiercely competitive information markets, this scenario does not necessarily result in the production of better quality information. On the other, critical, prompt, useful and quality information cannot be taken for granted just because new and more actors are participating in the shaping of information today. In this regard, that audiences have transformed into prodiences does not imply in itself that they are better equipped to participate in the generation of quality information.

At least in the near future, technological and commercial market trends will continue to exert strong pressures over media organizations and over journalistic practices within them. There, professional journalism will not disappear, of course, but profit and economic benefit will continue to lead the media market's logic. This is a reason why it is relevant not only to acknowledge the new roles that prodiences may be playing in the shaping of content, but also to think of such roles in a critical manner.

As we said before, today prodiences are able to receive, collect, report, interact, discuss, redefine, redesign, create and disseminate information. Nevertheless, their sole activity in the information production dynamic does not entail that their participation here may improve in itself the quality of content in terms of the needs of democratic society. Media literacy enters the stage here. Today, prodiences have new opportunities to shape content, but it must also be acknowledged that they have new "co-responsibilities" in the generation of debates, information and opinions in the public sphere. The preservation of an open society requires different types of free information exchange, from mere entertainment to in-depth debates, and perhaps also with different standards and characteristics. Media literate prodiences, as co-responsible factors for shaping information, represent a possibility for preserving such a free and wide landscape of information exchanges necessary for the safeguarding of an open and truly democratic society.

REFERENCES

Adams, D. & Hamm, M. (2001). *Literacy in a Multimedia Age*. Norwood, MA: Christopher Gordon.

Amaral, R. & Guimaraes, C. (1998). "Brazilian television: a rapid conversion to the New Order", in Elizabeth Fox (ed.), *Media and Politics in Latin America: The Struggle for Democracy*, Beverly Hills, CA: Sage.

Aufderheide, P. (1993). Aspen Media Literacy Conference Report Part II. Queenstown, MD: Aspen Institute. Available at www.medialiL.org/reading_room/article356.html/ Retrieved: October 6, 2010.

Bagdikian, B. (2004). *The New Media Monopoly*. Boston: Beacon Press.

Bellamy, R. & Palumbo, A. (eds.). (2010). *Political Accountability. Democracy, Accountability and Representation*. Cambridge, UK: Cambridge University Press.

Bowman, S. & Willis, C. (2003). *We the Media. How audiences are shaping the future of news and information*. The Media Center at the American Press Institute (Retrieved in its Spanish version, *Nosotros los medios: cómo las audierncias están modelando el futuro de las noticias y la información*. Edited by J.D. Lasica and translated by Guillermo Franco. At: http://issuu.com/merce9/docs/we_media_espanol Retrieved: September 17th, 2010.

Brunner, C. (1999). *The New Media Literacy Handbook*. New York: Anchor Books.

Chen, F. (2000). "Subsistence Crises, Managerial Corruption and Labour Protests in China" *China Journal* (44), 41-63.

Christ, W. & Potter, W.J. (1998). "Media Literacy, Media Education, and the Academy," Journal of Communication, 48(1), 5-15.

Curran, J. (1991). "Mass media and democracy: A reappraisal," in J. Curran and M. Gurevitch (eds.), *Mass Media and Society*, London: Edward Arnold.

Curran, J. (2002). *Media and Power*. London: Routledge.

Curran, J. & Seaton, J. (2003). *Power without Responsibility: The Press, Broadcasting and New Media in Britain*. Sixth Edition, London: Routledge.

Curry, J. L. (1989). *Poland's Journalists: Professionalism and Politics*. New York: Cambridge University Press.

Dalhgren, P. & Sparks, C. (eds.). (1997). *Communication and Citizenship: Journalism and the Public Sphere*. London: Routledge.

Datta-Ray, S. (1998). "Press freedom and professional standards in Asia," in A. Latif (Ed.) *Walking the Tightrope: Press Freedom and Professional Standards in Asia*. Singapore: AMIC.

Gans, H. (1979). *Deciding What's News*. New York: Vintage, 1979.

Gitlin, T. (1980). *Whole World Is Watching*. Berkeley: University of California Press.

Harris, P. (1988). "China: The Question of Public Service Accountability," in J.G. Jabbra & O.P. Dwivedi (Eds.), *Public Service Accountability: A Comparative Perspective*. Hartford, CT: Kumarian Press, 227-49

Heins, M. & Cho, C. (2003). *Media Literacy: An Alternative to Censorship*, 2nd (ed.) Free Expression Policy Project. Retrieved October 6, 2010. http://www.fepproject.org/policyreports/medialiteracy.pdf

Jabbra, J.G. & Dwivedi, O.P. (eds.). (1988). *Public Service Accountability: A Comparative Perspective*. Hartford, CT: Kumarian Press.

Johnson, O.V. (1998). "The media and democracy in Eastern Europe," in Patrick O'Neil (ed.), *Communicating Democracy. The Media and Political Transitions*. Boulder, CO: Lynne-Rienner Publishers.

Langer, J. (1998). *Tabloid Television: Popular Television and the "Other" News*. London: Routledge.

Lasica, J.D. (2003). "What is participatory journalism?" *Online Journalism Review*. August 7. Available at: http://www.ojr.org Retrieved: October 6, 2010.

McChesney, R. (2008). *The Political Economy of Media. Enduring Issues, Emerging Dilemmas*. New York: Monthly Review Press.

Meiklejohn, A. (1960). *Political Freedom: The Constitutional Powers of the People*. New York: Harper.

Michnik, A. and Rosen, J. (1997). "The media and democracy: a dialogue", *Journal of Democracy*, 8(4).

Mihailidis, P. (2009a). "Beyond cynicism: media literacy and civic learning outcomes in higher education," *International Journal of Learning and Media*, 1(3), 19-31.

Mihailidis, Paul. (2009b). "Connecting culture through global media literacy," *Afterimage: The Journal of Media Arts and Cultural Criticism*. 37/2, 37-40.

Mill, J.S. (1989). *On Liberty and Other Writings*, ed. by Stefan Collini, Cambridge Texts in the History of Political Thought. Cambridge, UK: Cambridge University Press.

Milton, J. (2008) [1644]. *Areopagitica: A speech of Mr. John Milton for the liberty of unlicensed printing to the Parliament of England*. Forgotten Books.

Murdock, G. (1990). "Redrawing the map of the communications industries: concentration and ownership in the era of privatization", in M. Ferguson, (ed.), *Public Communication*. London: Sage.

Nichols, J. & McChesney, R. (2010). *The Death and Life of American Journalism*. Philadelphia, PA: Nation Books.

OFCOM. (2005). "Media literacy audit. Report on media literacy amongst children", Office of Communication. London. Available at: http://www.ofcom.org.uk/advice/media_literacy/medlitpub/medlitpubrss/children/children.pdf Retrieved October 6, 2010.Popper, K. 1963 [1945]. *The Open Society and its Enemies*. New York/Evanston: Harper Torchbooks. Two Vols. 4th Edition.

Postman, N. (1985). *Amusing Ourselves to Death: Public Discourse in the Age of Show Business*. New York: Penguin Books.

Potter, J.W. (2004). *Theory of Media Literacy: A Cognitive Approach*. Thousand Oaks, CA: Sage.

Potter, J.W. (2010). *Media Literacy*. Fifth Edition, Thousand Oaks, CA: Sage.

Ritzer, G. & Jurgenson, N. (2008). *Producer, consumer...prosumer?* Paper presented at the annual meeting of the American Sociological Association, Boston.

Rivoltello, P.C. (2006). *Screen generation. Gli adolescenti e le prospettive dell'educazione*. Milano Vita, Pensiero.

Schedler, A. (1999). "Conceptualizing accountability", in A. Schedler, L. Diamond & M.F. Plattner (eds.). *The Self-Restraining State: Power and Accountability in New Democracies*. London: Lynne Rienner Publishers.

Shoemaker, P. & Reese, S. (1996). *Mediating the Message: Theories of Influence on Mass Media Content*. New York: Longman.

Siebert, F.S., Peterson, T. and Schram, W. (1956). *Four Theories of the Press. The Authoritarian, Libertarian, Social Responsibility and Soviet Communist Concepts of What the Press Should Be and Do*. Freeport, NY: Books for Libraries Press.

Silcock, B.W. & Keith, S. (2006). "Translating the Tower of Babel? Issues of definition, language and cultura in converged newsrooms," *Journalism Studies*, 7(4), 610-627.

Silverblatt, A. (2001). *Media Literacy: Keys to Interpreting Media Messages*. 2nd Edition. Westport, CT: Praeger.

Singer, A. (1994). "Nota sobre o papel da imprensa na transicao brasileira", in C. H. Filgueira & D. Nohlen, (eds.), *Prensa y transición democrática: experiencias recientes en Europa y América Latina*. Frankfurt an Main: Vervuert.

The Random House Dictionary of the English Language. (1987). Second Edition. New York: Random House.

Thoman, E. (2010). "Three Stages of Media Literacy" from *The Media Awareness Network*. Retrieved October 6, 2010. http://www.media-awareness.ca/

Tracey, M. (1988). *The Decline and Fall of Public Service Broadcasting.* Oxford, UK: Oxford University Press.

Tuchman, G. (1978). *Making News.* New York: The Free Press.

Tumber, H. (2001). "Democracy in the information age: The role of the fourth estate in cyberspace". *Information, Communication & Society,*
http://www.informaworld.com/smpp/title~db=all~content=t713699183~tab=issueslist~branches=-v4 (1))March 95-112.

Ungar, S. (1990). "The role of a free press in strengthening democracy ", in J. Lichtenberg (ed.), *Democracy and the Mass Media.* Cambridge, UK: Cambridge University Press.

White, D.M. (1950). "The gatekeeper? A case study in the selection of news," *Journalism Quarterly,* 27 (74): 383-390.

Zavoina, S. & Reichert, T. (2000). "Media convergence/management change: The evolving workflow for visual journalists," *The Journal of Media Economics,* 72(2), 143-151.

Chapter 3 -
Global News Literacy:
Challenges for the Educator

STEPHEN D. REESE
University of Texas at Austin, USA

Introduction: Finding the Global in Global News Literacy

Educators face the same shifting landscape of global news as do professional practitioners, citizens, and media scholars. The rapid changes in technology have given rise to new media platforms and greater interconnectedness while dramatically altering traditional news institutions and eroding professional boundaries. This raises new questions about the potential for cross-cultural understanding and the values of cosmopolitan citizenship. This interconnectedness is one of the hallmarks of globalization, which along with a simultaneity and synchronization of communication contributes to our impression of the world as a single place. These networks of international journalism support what I've called a "global news arena" (Reese, 2008), which brings about pressures toward transparency, both on the part of governments and from journalism. Slanted or false reports are now more rapidly challenged or augmented—not only by other news organizations but by thousands of readers and viewers who circulate and compare reports through on-line communities. Old criticism of news "bias," although still fervently expressed in the U.S. political debate, are now more multi-layered and carried on communication platforms that themselves cut across national boundaries. The popular 2011 uprisings in Tunisia, Egypt, and throughout the Middle East, were facilitated by Internet communication, even when the regimes tried to regulate traffic outside the country. The Qatar-based Al-Jazeera television news service provided some of the best coverage of the Egyptian revolt, but its availability in the U.S. was limited by cable operators failing to provide it to their subscribers. In spite of being deemed anti-American by some U.S. critics, the value of its coverage in a critical world hot-spot gave it new professional prestige and led to heightened demand for internet streaming of its programming.

International communication long has been a research subject in many academic programs, introducing students to various media systems from a comparative perspective. And certainly global news literacy requires a basic awareness of how national media contexts differ. This kind of cross-national comparison is firmly embedded in the rankings of countries in, for example, their relative levels of press freedom. Beyond *within*-country descriptions, other analyses have examined how news flowed *among* these countries, and the relative imbalances in that transmission, producing distorted images both within and across countries. World news coverage was often limited to the developed world, only paying attention to the "third world" when something bad happened, commonly during political unrest or large scale natural disasters. The dominant Western news agencies were seen as exerting "hegemonic" power over the world's news images, causing developing countries to receive even news about themselves through non-indigenous professional filters. More recent research, however, has tried to capture the complex inter-relationships at the subnational, extra-national, and transnational level that characterize globalization.

In many ways, pedagogical trends have tracked these changes in the communication world and the related research. As global news becomes more complex and multi-layered, however, the subject challenges even the most conscientious instructor to cover the material adequately and with sensitivity toward other national contexts. In this respect, news literacy is not unlike many subjects where the explosion of knowledge makes "covering the material" impossible. Even those courses and texts that are sensitive to the broader global phenomena still often seem limited to presenting an inventory of media phenomena across a range of countries. True comparative analysis is rare. Volumes in the area of the growing field of globalization studies itself often treat media as an afterthought, regarding CNN, for example, as an exemplar of "global media." Global, however, is more than something really big, beyond the national levels, or something that simply happens in some other country. A global perspective also means taking concrete local circumstances into account while being aware of how they differ from other areas, and how global forces bring "influence from a distance."

If we are to take news literacy in a global direction, then, it must go beyond traditional critiques of international news. Citizens are able as never before to be reflexive about news, given their access to alternative sources and perspectives, making interest in news literacy a worldwide phenomenon. The likelihood of citizens becoming news literate, however, is not a foregone conclusion. News is still domesticated through national frames of references, often taken for granted, and media globalization skeptics have argued that no truly transnational news platforms have emerged, permitting the kind of cross-boundary dialogs associated with a public sphere (e.g., Sparks, 2007). Indeed, critics like Hafez (2007) suggest that regional enclaves, linguistic divides, and parochial zones of ethnocentric discourse have become even more sharply defined. Such skeptics point to the continued weaknesses of international reporting: elite-focused, conflict-based, and driven by scandal and the sensational, leading them to conclude that the "global village" has been blocked by domestication (reviewed in Reese, 2010). Even in the U.S., where the press system is advanced and highly professionalized, elite journalists reinforced the discursive echo-chamber supporting the decision to go to war in Iraq by internalizing the War on Terror frame promoted by the Bush administration (Reese & Lewis, 2009).

Thus, news literacy, even in the midst of more journalistic sources than ever before generated by both citizens and professionals, must be taught and cultivated. By news literacy I essentially mean an understanding of how news "works," including the underlying media and technological systems that support certain meanings embedded in media "texts" and the creative process that yields them. Although an incredibly loaded term, "literacy" is a social practice, and that means locating it within a set of power relations while considering its moral, political, and cultural context.[1] Global news literacy, then, means the ability to understand, "decode," and create media with particular awareness of one's social location within an international context. We need not only a deep understanding of how news is working in certain particular cases but also to understand how to connect these cases to a larger framework. In teaching, in spite of the complex and abstract-sounding concepts within globalization, a case-oriented approach is often helpful for such purposes. In this chapter I consider some ways educators can adapt to the teaching of news literacy in a global context, drawing from my own analysis of the globalization of journalists (Reese, 2010). And I will note some approaches in my own teaching experience, both at the University of Texas and the Salzburg Academy on Media

---

1 For a review of recent work in media literacy education and research see the volume by my University of Texas colleague, Kathleen Tyner (2010).

& Global Change,[2] that seem relevant to this challenge. We obviously want to approach news literacy with full consideration to the global context, and find appropriate instructional strategies, but how do to this is often a challenge.

Media Literacy Teaching Trends

The media literacy movement, examined elsewhere in this volume, intersects with a number of trends in education, and provides a fruitful way to tackle the issues raised by media globalization. The educator must consider how the theoretical insights from media research map onto the pedagogical strategies available under the umbrella of media literacy. Indeed, the changes in media facilitate the very advances in teaching advocated by so many: critical thinking, active experiential learning, writing across the curriculum, and collaborative inquiry.

For the educator, these pedagogical trends have been supported in large part by changes in technology. One of the main strengths of the web is that it can help foster individualized learning. Kozma and Johnston (1991) acknowledged this early on in the web's development, observing that "the computer's processing capability can be used to create procedural systems in which information provided by the user determines what happens next" (12). Such an individualized approach has changed the role of the instructor in the learning process. Branson (cited in Menges, 1994) argued that this shift away from the professor as the center of the classroom was part of a new paradigm for education, with a new center occupied by a collective, "accumulated knowledge" to which the students (as well as the professor) have direct access. (This models today's wider on-line communities.) Students learn through interaction with peers, with professors and through the use of new technology. Internet connectivity allows students to discuss their research with other students from

---

2 The Salzburg Academy on Media and Global Change, mentioned throughout this text, has worked over the past four years to build pedagogical models for media and news literacy that address global audiences. For three weeks every summer, more than fifty students and a dozen faculty from fifteen universities worldwide gather to explore media's role in global citizenship and civil society. The primary outcome of the Salzburg Academy is a student-created curriculum on Global News Literacy. This curriculum is founded on the notion that global citizenship and responsibility require individuals to have an understanding of media's necessary role in society and an awareness of the ways in which media influences cultural ideologies both locally and globally. Through the creation of dynamic educational content that investigates media's role in global society, the participants at the Salzburg Academy enter into cross-cultural dialogue that, at its core, reflects new understandings of media from diverse perspectives.

different geographic regions, ethnic groups, government systems and econo-
mies. The combination of a multiplicity of resources and viewpoints pushes
students to consider "facts" as presented by media as transitory and problematic rather than as static and opaque as textbook information. This is a perfect
illustration from a technological standpoint of the teaching style called for in
conjunction with news literacy. As Kozma and Johnston (1991) note, "with
technology, students are moving away from the passive reception of information to active engagement in the construction of knowledge" (16).

Media literacy can be regarded as a subset of critical thinking, which according to the Foundation for Critical Thinking on its website is defined as
"the art of analyzing and evaluating thinking with a view to improving it." To
the extent that it involves questioning, reasoning, discerning the strength of
claims, evaluating evidence, and taking multiple perspectives, media literacy
necessarily is thinking critically, a process we presume leads to better informed
citizens, who can evaluate the strength of political arguments and detect faulty
logic as they make decisions. The public sphere concept of Jürgen Habermas
emphasizes this thinking by posing a normative ideal of a discursive space,
widely accessible to ideas that compete on the basis of reason. In news literacy,
we wish to promote a pedagogy of inquiry, to make "asking critical questions
about what you watch, see, and read" stand at the center of what it means to
be media literate (Hobbs, 1998, 28). Thus, media literacy shares with critical
thinking initiatives the promotion of intellectual autonomy on the part of the
student/citizen. The broader goal of critical thinking guards against taking the
mediated environment for granted. We want people to be able to stand back
from news media objects, aesthetically, politically, and intellectually.

Looking back, it seems the world has caught up with media literacy, and
many of its concerns now have become familiar issues. The concept of "literacy" itself seems straightforward, people should be able to "read" and "write."
Applying this idea to a broader array of media, beyond the traditional printed
word, signals that they are also important and that the messages they contain
(particularly visual) are not altogether obvious. Daily media consumption, although often deemed less important in the past as "mass" or "popular" culture, carries its own invisible curriculum along with more highbrow texts that
are studied in the canon. Young people were considered at risk from the
power of media, particularly television in the early years, and needed some self-
defense tools. As university professional schools in communication have
grown steadily these skills have become institutionalized, rooted in the various

departments training students to analyze media texts, understand their social context and power, and produce media themselves. Digital media, now small and cheap enough to be widely available, have made this participatory part easier, and in doing so brought a heightened critical awareness to the other areas as well. Media literacy always had at least an implicit political agenda, often liberal but seldom radical. In the early going, do-it-yourself media projects could be regarded as a mild reformist impulse, holding out an emancipatory potential for the student-citizen-community producers while drawing attention away from the more deleterious media effects and leaving the entrenched commercial and corporate basis for the media intact. Now, the spread of citizen journalism available across the political spectrum has not only vastly increased the volume of media critique but itself encroached on these media structures by redrawing professional boundaries (e.g., Allan & Thorsen, 2009).

Global Thinking/Global Teaching

Just as in media literacy we are led to no longer take media for granted and in critical thinking to no longer take the thinking process for granted, so does globalization itself bring greater reflexive awareness of ourselves in a larger network. We have more with which to compare ourselves and do so in real time—thus, problematizing the local. Becoming aware of the world as a "single place," an intuitively appealing marker of globalization, causes us to stand in ironic detachment to ourselves. Such globalization insights help guide instructional strategies. In particular, I have observed and would like to suggest that using case-oriented instruction, at the Salzburg Academy adapted to a lesson-plan technique (described in Chapter 5 of this volume), encourages a useful teaching culture where so many perspectives intersect. It mirrors in some respects the theoretical distinction between the global and the local.

The "global village" perspective regards the global as one within a nested hierarchy of levels of analysis based on size: the global then lies beyond the local, regional, and national. Against this expectation that media report and reach the entire globe, however, little evidence exists for a world communication system with an undistorted view of the world. The global village also implies global consciousness, which implies a homogeneity of world views, or at least a diverse "dialog of cultures." Again, the global media system, particularly international broadcasting, doesn't live up to that hope: homogenization loses out to domestication. The "networked society" theorizing of Manuel Castells

(Castells, 2007), for example, is interpreted by some to necessarily require a giant cluster of inter-linked world, state and cultural entities, but it should rather be seen as yielding different lines of cross-border articulation. Rather than assume that globalization means a uniform imposition of a global (village) standard across a range of local circumstances, we should more realistically consider it as a complex interplay between local and cultural forces from a distance.

Satellite news channels have figured prominently in the "media globalization" debate. This has led to these platforms often being regarded as a "space apart" in a new "global" realm. Volkmer (1999), for example, tied global news to an emerging world civil society structure. In her study of CNN International, she argued that global political communication constructs a global public sphere, from which emerges global civil society. She further argues that the global public sphere is a new political space, with the capacity to pressure national politics and provide communication not otherwise possible on a national level—with, for example, the Arab television network Al-Jazeera's interview with Osama bin Laden, or extra-territorial websites set up by Chinese dissidents (Volkmer, 2002). No doubt that's true, but calling it a new "sphere of mediation" can also be misleading.

This globalization theorizing implies that there is some global space apart from the local, which led Robertson (1995) to propose the helpful concept of the "glocal." Observing the globalization process means finding where the global intersects with the local, where the universal meets the particular. This is helpful because when it comes to media globalization, many are inclined to expect a special communication platform that serves as a virtual global space. But the global public sphere is not some new autonomous zone, operating like national public spheres only on a broader supranational scale (most obviously in zones like the European Union). When globalization "skeptics" define it that way, of course they are led to deny its existence. Hjarvard (2001), rather, claims it is a process of restructuring and recasting public communication. Globalization of the public sphere means the *process* by which the national spheres become deterritorialized, not the creation of a new and separate global public sphere but a "multi-layered structure of publicity" (34). As appealing (and utopian) as the concept may seem, there is no "global" public sphere per se, floating free of localities, and attempts to theorize one break down in the absence of a more defined and observable social space.

This raises the question then of where to go to observe these relationships and how to teach them. We are embedded in communities beyond the ones we live in, ones not defined by place. That doesn't mean, however, that physical place has ceased to matter for global level processes. The work of globalization theorists in geography and sociology leads us to seek the workings of the global in specific local places, where the universal and global become particularized and local. Global networks don't exist virtually but connect local nodes, where people interact in real places with key members of other networks, and where they develop common norms and logics necessary for the functioning of complex global exchanges. Thus, Castells' network society (2007) embraces the "space of flows" and the "space of places" and their interactions, capturing the seeming paradox that in the face of so much on-line communication such global cities as London and New York have become even more important, where the global and local come together in the interaction of cosmopolitan elites.

Teaching global news literacy means following these connections, not simply laying claim to CNN or Al-Jazeera as themselves prima facie evidence of global influence. This is a challenging task for the educator, but encouraging in the sense that the global can be tackled through an analysis of concrete cases in local circumstances. No matter where students are from, they can understand the global through its connection with a specific familiar local context. They cannot think about it only in the abstract, but must find those insights through the concrete particulars.

Global News Literacy Case: China

A global news literacy perspective is particularly helpful in understanding the complex process of social change in countries with highly controlled press systems now being witnessed in a number of regions. That goes both for those in the middle of such changes and onlookers like myself. My own scholarly understanding of the globalization process has been enhanced by working as an educator with insiders via the Salzburg Academy, but this has been a mutually influential process in a way that illustrates some of the unexpected kinds of learning that a focus on news literacy can bring about.

In understanding global news processes, it is not always helpful to divide countries into free and unfree, but rather we should consider the introduction of new "spaces" for public deliberation that are made possible by global interconnectivity and communication technology (Reese, 2009). The remarkable online changes in China, with the world's largest group of "netizens," have also brought about unexpected deliberative openings in the public sphere. The rising number of "mass incidents" fueled by the Internet brings pressure to bear on government corruption and other public concerns. The government still has many tools for controlling communication, but these changes are moving inexorably toward greater societal openness. Even in one of the most carefully stage-managed events, the Beijing Olympics, the director of the Asian Studies at Georgetown, Victor Cha, claimed that social change had been brought about in one of the world's most rigid systems, a seismic shift that cannot be undone (Cha, 2008). The Chinese netizens have shown how media criticism, or "news literacy," can go viral with significant new consequences. Driven by nationalism, the government encouraged criticism of foreign press performance in covering China, but the tools of news literacy are not so easily confined to external targets and can easily be redirected toward more homegrown media problems. These tools include the ability to evaluate, critically analyze and compare media portrayals, as supported by a variety of digital platforms, blogs, bulletin-board systems (BBS), forums in traditional news sites, and social media. Thus, rather than see social change as the result of a "McDonaldization" process of world homogenization, a cultural globalization process is taking place not in a directed top-down imposition of force but through spaces of mutual awareness, in which standards evolve in a reflexive process (Robertson, 1995).

I understood this dynamic better after hearing firsthand from Chinese colleagues. In teaching at the Salzburg Academy in 2008, I heard a presentation by the professor and delegation of students from the Chinese partner university on the international press coverage leading up to the Beijing Olympics. One of the issues at the time concerned how foreign news reports were depicting social unrest in Tibet, and CNN had been a particular target, serving as exemplar for other Western media. CNN commentator Jack Cafferty inflamed this sensitivity by calling the Chinese government, "the same bunch of goons and thugs they have been in the past fifty years." The Chinese amateur website, Anti-CNN.com, was established to help collect what were considered lies and distortions, aided by thousands of contributors, or "netizens," in a process sometimes called "human flesh search engine." The most remarkable action taken was the establishment of online petitions against the Western media (especially CNN) which reportedly accumulated tens of thousands of signatures and comments. Netizens wanted Western media to respond to their petition, stop publishing irresponsible comments, and apologize for biased news coverage.

The Chinese group's analysis at the Academy reflected their more general concern toward how the country would be viewed on the world stage, and some images were shown illustrating errors and distorted emphasis in media coverage. Although motivated by national sensitivities, it struck me that the Chinese were discovering "news literacy" techniques and applying them on a scale never seen before. The problems of ethnocentrism, news sensationalism, dramatic visuals, reporter pack mentality, lack of historical context, reliance on a narrow group of sources, and distorted emphasis were not peculiar to coverage of China. They are tendencies of news media more generally and the subject of numerous Western-based research critiques. I was led to do my own analysis (an essay I shared with the instructor and later published) of how that same enthusiasm for media criticism could easily be turned inward after Olympics fever subsided, giving rise to greater demands for media accountability inside China (Reese & Dai, 2009). The ever increasing number of internet-amplified "mass incidents," many of which have included incident of journalistic malpractice, suggest this is the case.

The following year another delegation of students came to Salzburg with a different professor, and I was intrigued to see a new presentation on how Chinese netizens, as citizen journalists, were monitoring and helping change incidents of corruption and violation of social norms. For example, a local official was seen in one image posted on the Internet wearing an expensive watch, one not affordable at the salary he was paid in his government post. In another incident, an official was smoking a similarly pricey cigarette. Ferreting out and commenting on such images spurred investigations into local corruption. Like the army of bloggers in the U.S. and elsewhere, Chinese citizens were scrutinizing information available on the Internet in the interest of better government. I was particularly interested that the students had quoted me in their presentation making reference (in the chapter above) to a telling quote I had found provocative from the *New York Times*–that a revolution was coming to China, only that the revolution is experienced "mostly as one of self-actualization: empowerment in a thousand tiny, everyday ways" (Thompson, 2006, 64). A case-style approach to news literacy, which I take both years' presentations to be, had yielded valuable insights in being seeded across national contexts.

News Literacy Instructional Directions

A number of related pedagogical initiatives in recent years, with which the author has been involved, have spoken to this problem of how best to engage with a globalized news media. They don't stem specifically from a media literacy context per se, but they do relate to the same objectives of how to get students to be more intellectually rigorous in their relationship with media. The common thread for three of these initiatives has been the Knight Foundation, a philanthropic power with roots in the newspaper industry. Knight shared our view in journalism education that the best way to improve news literacy is through the schools, laying the groundwork early for informed citizens. That fits with the somewhat evangelical quality of the media literacy movement, which seeks to disseminate media understanding through the schools, and that means taking pedagogical strategy seriously. In two Texas projects we aimed to teach teachers as a way to seed valuable ideas in the classroom, organizing the projects for participants around having them construct lesson plans around what they had learned. The Salzburg Academy has taken a similar approach in compiling lesson plans from participants, using a case or story to introduce broader principles.

Media & American Democracy Project

Teachers must increasingly engage with students as co-investigators. Rather than a repository of information that students tap for knowledge, the teacher guides them in their own discovery. This can feel like giving up control and diminishing the leadership role, but it requires its own special skill. It's also a constant process of balancing structure with freedom.[3] Between 1997 and 2000, the Media and American Democracy Project brought the Harvard Graduate School of Education, Joan Shorenstein Center on the Press, Politics and Public Policy, and the Kennedy School of Government together to train high school teachers in a multi-day institute. The goal was to expose teachers to the latest ideas and research related to the press, politics, and American policy: how media are used in the democratic process, and how new media

---

3 In one of my Salzburg experiences, the U.S. students were asked, apart from the other curriculum, to prepare a presentation about the American media system to share with everyone else and gathered to discuss how to proceed. As a faculty leader I was inclined to give them wide latitude and let them "self-organize" (after all, they should be the experts in the U.S. media), but after an hour of chaotic discussion they clamored for some guidance and structure (at which point I dispatched my wife, Carol, a master middle-school teacher!).

technologies affect the political process. In this respect the emphasis was learning "about" media and their institutional importance in the political system, but when we partnered with them to carry out a Texas program we added the goal of teaching students to be "savvy and critical news consumers." As the professional news industry has changed dramatically in the last decade it has only increased the importance of schools of journalism to pay as much attention to citizen news literacy as to training industry practitioners.[4]

This project had a strong pedagogical dimension in showing how to develop course instructional methods centered on media topics. At the heart of this approach was the Teaching for Understanding Framework, developed by Howard Gardner and others at Harvard. The key from a news literacy perspective was the focus on a spirit of investigation. Teachers developed curriculum units using Generative Topics (What is the subject? E.g., some news media issue), Generative Objects (What object(s) evoke images central to the focus of the unit?: e.g., a news story or image), Understanding Goals (Essential Questions and Critical Engagement Questions), and Performance of Understanding. The rationale for this approach is to provide intellectual coherence to what is being learned, so that the student is pushed toward understanding beyond basic knowledge of facts (information on tap). Teachers develop essential questions to which the material contains the answers: "What makes news?" "Is it possible or desirable to have 'unbiased' news?" "To what extent are journalists free to write what they please?" Such questions suggest investigation beyond the more rigid and directive term "objectives," which imply preset answers. Such guiding questions are better able to lead to understanding, where one can think and act flexibly with what is known.

A framework with this kind of "optimal ambiguity," according to David Perkins of Harvard's educational Project Zero, provides enough structure and flexibility to meet the needs of classroom teachers, who are encouraged to personalized their innovations. I suspect teachers may have been doing similar things before, but having the imprimatur of Harvard helped certify its value and give them permission to be more open-ended in their approach and less a slave to having to "cover the material"[5] (Blythe & Perkins, 1998).

---

4 With support from Knight and Carnegie Foundations my school has launched new courses in recent years intended to introduce these concepts to the wider campus.

5 I suspect there is a countervailing pressure among public school teachers who have in recent years been subjected to increasing assessment measuring and the need to "teach to the test."

High School Journalism Institute

A similar approach was taken when we designed an institute for high school journalism teachers, in a project for the American Society of Newspaper Editors (also sponsored by Knight) designed to revitalize secondary school journalism (and in turn, it was hoped, the profession). We asked the participants to design curriculum units as part of their learning process, and these have been posted on the organization's website over the last several years. Using the same Teaching for Understanding Framework as with the Harvard project, we directed them to develop these lesson plans around journalism issues but based on finding key generative objects and essential questions, from which broader discussions and student projects could grow. Of course, when it comes to news there is no end of specific instances of coverage that can serve a jumping off point. Without being too strict with pedagogical theory, the lesson project provided a way for these teachers to adapt what they learned to their own needs, developing specific and practical classroom resources that would help them when they returned to their campuses.[6] As our Salzburg students have developed their own lessons plans, we imagine that they will have a similar sense that their work has been organized around issues of concern to them and motivated by the expectation that it will be publicly available.

Critical Thinking/Case Studies

In my own teaching I have been influenced by these initiatives, particularly in cultivating a style of questions in the classroom that helps with critical engagement. In that respect, I find intersection between the Teaching for Understanding approach and case studies. In both, one can start with specific facts (or generative object) and from there in a Socratic process develop guiding questions that bring out larger principles and deeper understanding.

In my school of journalism, we experimented several years ago with a new class called "Critical Thinking for Journalists." Given the importance of clear thinking for aspiring journalists we had high hopes for a class where thinking itself was at issue, with significant attention to informal logic, fallacies, and multiple perspectives. As the instructor I soon saw the point of one of the criticisms of the critical thinking movement, that it's too self-referential and become too sealed off from substantive issues. Not doubting the value of sound reasoning,

College level teacher, although increasingly obliged to implement their own assessment culture, are relatively more free in their instructional style.
6 See http://www.hsj.org/modules/lesson_plans/ for numerous examples of such lessons.

can we think about thinking itself or must we always be thinking about something? As enrollment grew it became impractical to teach those skills in a large format courses. Thus, in due time the course evolved into something more general, "Critical *Issues* in Journalism," but I retained an interest in getting students to think more critically about their future profession.

After several years away from teaching that course, I incorporated upon returning a case study approach. Within journalism education, reform efforts were underway to introduce the same kind of reflective teaching methods more common in other professional schools like law and business, long accustomed to using cases. This was part of a more general movement from inside the profession to recuperate, in the face of stagnation and ethical drift, the essential qualities of journalism: Project for Excellence in Journalism. A few years later, a special initiative was launched in 2007 with support from the Knight Foundation and housed at Columbia's School of Journalism, with the goal of developing high quality case studies and encouraging faculty to consider them as a way of cultivating critical thinking, through the lens of the professional, to train "reflective practitioners" (Schoen, 1983). Journalism educators, like others, are not quick to change their teaching culture, and so the case method, which requires some significant teacher retooling, has not yet been widely embraced. According to its promotional materials, "The goal is to train students to think like managers and leaders. Case-based courses develop students' analytic, decision-making, management and leadership skills. Through class discussion, students have the opportunity to examine in depth a wide range of editorial, ethical and economic issues. Students have a chance to exercise the muscle of judgment, strengthening it for the time when they will have to use it in a newsroom or other professional setting."

I have used a collection of textbook cases, including classic journalistic moments such as Watergate, the Clinton/Lewinsky scandal, and the Columbine school shooting tragedy (Rosenstiel & Mitchell, 2003), supplementing them with more recent cases from the Columbia collection, including one that explored, for example, the ethics of reporting the presumed sexual orientation of a U.S. soldier killed in Iraq. In using these cases as the basis for the course, it is easy to introduce critical thinking skills into classroom discussion: how well justified was the decision, what other perspectives are possible, why did the journalist decide as he or she did and what were the ethical implications? When I have students develop their own case studies, they learn to better discern the "professional issue" from a set of facts. For many of them, having only recently left high school, this is often a challenge.[7]

The case study approach lines up with the Salzburg-Harvard-ASNE lesson-plan strategy in allowing students to begin with specific concrete circumstances (generative objects) and build from there to broader understandings. In the case of news literacy subjects, they start from something accessible and build out from there. In my first experience with the Salzburg Academy in 2008, I could see that the diversity of students and national perspectives made something like this teaching approach necessary. Although it is helpful for students to have a common grounding in certain knowledge, there is no practical way to "cover the material" in a subject so broad as "global media," especially when handling a group coming to the table with such different levels of prior knowledge. But the range of generative objects in the form of "stories" that they begin with is wide, and with the media tools at their disposal they can take the information they acquire and develop their own "performances" to deepen their understanding.

Conclusion

In teaching global news literacy educators face a complex but engaging subject, but have at their disposal a array of teaching strategies ideally suited to the task. By engaging students in their own deep reflection about news media issues, using materials particular to their interest and cultural setting, this approach is readily adaptable to any cultural context. It engages students, facilitates active learning, and gives them greater appreciation of how their experience fits into the larger framework of global news. I have considered some of these principles and tied them to the pedagogical process as used in the Salzburg Global Academy and my own institution at the University of Texas. In my reference to specific projects, I hope in this admittedly personal narrative to have shown how these approaches relate to the challenges of teaching news literacy, especially in a global context.

If it were ever possible to "cover the material" of any classroom subject in the past, explosion of knowledge has made that day long past. Instead there are habits of mind, intellectual self-defense skills, that can be taught in ways that we hope are engaging and lead to more self-directed learning. In the case

---

7 A range of news stories and commentary that often serve as the basis for case studies can be found on my occasional class blog: "press conference," http://sdreese.wordpress.com/.

of news literacy, we can emphasize understanding growing out of specific cases and exploit students' already easy familiarity with the ubiquitous media in their lives to use it as a window into broader issues. Although our scope is global we can still approach news literacy through local frames of reference; indeed, how could we do otherwise? As considered in my examples above, using lesson plans projects, as in the Harvard and ASNE initiatives, or the case studies that are becoming more integral to journalism education oblige students to learn what is necessary in order to answer the fundamental questions about how they live their lives both *in* and *through* media. The Academy model, itself, may not be possible everywhere, but it shows how a single subject (as in, for example, my China case) can be engaged by a diverse set of students bringing their own unique perspective and building a model for news literacy understanding. The juxtaposition of perspectives brings about richer insights, which are particularly needed as we confront the complex web of media and citizen connections in a more globalized world.

REFERENCES

Allan, S., & Thorsen, E. (Eds.). (2009). *Citizen Journalism: Global Perspectives*. New York: Peter Lang.

Blythe, T., & Perkins, D. (1998). The Teaching for Understanding Framework. In T. Blythe (Ed.), *The Teaching for Understanding Guide*. San Francisco: Jossey-Bass, 17-24.

Castells, M. (2007). "Communication, Power and Counter-Power in the Network Society." *International Journal of Communication*, 1, 238-266.

Cha, V. D. (2008, August 4). "Beijing's Catch-22." *International Herald Tribune*.

Hafez, K. (2007). *The Myth of Media Globalization*. Malden, MA: Polity.

Hjarvard, S. (2001). News Media and the Globalization of the Public Sphere. In *News Media in a Globalized Society*. Goteborg, Sweden: Nordicom, 17-40.

Hobbs, R. (1998). "The Seven Great Debates in the Media Literacy Movement." *Journal of Communication*, 48(1), 16-32.

Kozma, R., & Johnston, J. (1991). "The Technological Revolution Comes to the Classroom." *Change*, 23(1), 10-23.

Menges, R. (1994). Teaching in the Age of Electronic Information. In W. McKeachie (Ed.), *Teaching Tips: Strategies, Research and Theory for College and University Teachers*. Lexington, MA: D.C. Heath & Co.

Reese, S. D. (2008). Theorizing a Globalized Journalism. In M. Loeffelholz & D. Weaver (Eds.), *Global Journalism Research: Theories, Methods, Findings, Future*. London: Blackwell, 240-252.

Reese, S. D. (2009). "The Future of Journalism in Emerging Deliberative Space." *Journalism: Theory, Practice, Criticism*, 10(3), 362-364.

Reese, S. D. (2010). Journalism and Globalization. *Sociology Compass, in press*.

Reese, S. D., & Dai, J. (2009). Citizen Journalism in the Global News Arena: China's New Media Critics. In S. Allan & E. Thorsen (Eds.), *Citizen Journalism: Global Perspectives*. New York: Peter Lang, 221-231.

Reese, S. D., & Lewis, S. (2009). "Framing the War on Terror: Internalization of Policy by the U.S. Press." *Journalism: Theory, Practice, Criticism*.

Robertson, R. (1995). Glocalization: Time-Space and Homogeneity-Heterogeneity. In M. Featherstone, S. Lash & R. Robertson (Eds.), *Global Modernities*. London: Sage, 25-44.

Rosenstiel, T., & Mitchell, A. (Eds.). (2003). *Thinking Clearly: Cases in Journalistic Decision-Making*. New York: Columbia University Press.

Schoen, D. A. (1983). *The Reflective Practitioner: How Professionals Think in Action*. London: Temple Smith.

Sparks, C. (2007). "What's Wrong with Globalization?" *Global Media and Communication, 3*(2), 133-155.

Thompson, C. (2006, April 23). Google's China Problem (and China's Google Problem, *The New York Times*.

Tyner, K. (Ed.). (2010). *Media Literacy: New Agendas in Communication*. New York: Routledge.

Volkmer, I. (1999). *News in the Global Sphere: A Study of CNN and Its Impact on Global Communication*. Luton, UK: University of Luton Press.

Volkmer, I. (2002). Journalism and Political Crises. In B. Zelizer & S. Allan (Eds.), *Journalism after September 11*. New York: Routledge, 235-246.

Chapter 4 -
Reaffirming the "Journalist" as Vital to 21st Century Information Flow, Civic Dialog, and News Literacy

RAQUEL SAN MARTÍN

Pontificia Catholic University, Argentina

Introduction

In July 2010, *Jornal do Brasil*—one of Brazil's oldest and most prestigious news-papers—announced that it would permanently close its print edition. Citing "financial difficulties" and a declining circulation, *Jornal do Brasil* moved its operation entirely online. Around the same time, Argentinean Foreign Affairs Minister Héctor Timerman restricted his communication with the media to just one platform: Twitter. Years earlier, in June of 2005, the *Los Angeles Times* launched *wikitorial*, an interactive platform allowing readers to contribute to and rewrite its editorial column, in what was called a "communal search for truth." Two days later, the experiment was aborted "because a few readers were flooding the site with inappropriate material," explained the newspaper to its online readers. In January 2010, after the devastating earthquake in Haiti, the social networking sites Facebook and Twitter showed the world the first images of the devastation, delivered SOS calls and organized international aid campaigns.

These examples help illustrate the turbulence and uncertainty affecting journalism in the hypermedia age. Declining revenues, shrinking audiences, and the vast growth of new media are together reshaping the landscape for journalism and news. It comes as no surprise, then, that journalism has been seemingly immersed in a professional identity crisis. Scholars and practitioners have diagnosed a variety of ailments for journalism, among them are: loss of credibility, rising public distrust, low salaries and poor working conditions, lack of adequate training, increased focus on efficiency and profit instead of quality, and the feeling that journalism is failing to interest people in the news.

At the heart of this so-called crisis is technology. Specifically, the emergence of social media and mobile technologies has revolutionized information and news flow carrying with them undoubted consequences for both newsmakers and audiences.

On the one hand, the impact of digital media platforms on news creation, distribution and reception is not only changing the news industry but also public attitudes towards media. The digital news world allows the public to highly individualize their news consumption and react to what is published immediately. This new reality is putting traditional news business models to the test, and reforming the daily rituals and functions of newsrooms all over the world. On the other hand, from the practitioners' point of view, digital media are questioning the culture that has shaped professional journalism for decades. Objectivity, fairness, accuracy, credibility, and the journalist as watchdog, are all being carefully reconsidered in the current climate of information.

As a result, a new media ecosystem is emerging: one that is increasingly integrating professional and amateur approaches to information gathering. This new environment for news is forcing journalists to address matters that were once beyond their purview. In this sense, the ubiquity of information has shifted the public's role in the news process. While some audiences are increasingly engaged in civic issues through new social media platforms, others seem to have become polarized through the selective filtering of information and opinion, and still others remain isolated from the digital world.

Nevertheless, there is reason for hope. First, far from shrinking, news media are growing. Second, the new spreadable nature of information is forcing journalists to build agendas that must satisfy both local and global audiences. These developments beg the question: how can future journalists cultivate relevant approaches to reporting in an increasingly global and hypermedia 21st century landscape. The answer lies in fostering new approaches to reporting that involve the public, mobile platforms, and the tenets of news literacy.

Journalism at a Global Crossroads

Over the last decade, the more pessimistic diagnoses of the state of journalism highlight a decrease in the quality of news coverage and a growing distance between the newsroom and the public. Evidenced by the "State of the News

Media," report by *The Pew Research Center's Project for Excellence in Journalism* (2010), the journalism industry's traditional models are indeed breaking down:

> "The old model of journalism involved news organizations taking revenue from one social transaction –the selling of real estate, car and groceries or job hunting- and using it to monitor civic life –covering city councils and zoning commissions and conducting watchdog investigations. Editors assembled a wide range of news, but the popularity of each story was subordinated to the value and the aggregate audience. And the value of the story might be found in its consequence rather than its popularity. That model is breaking down. Online, it is becoming increasingly clear, consumers are not seeking out news organizations for their full news agenda. They are hunting news by topic and by event and grazing across multiple outlets. This is changing both the finance and the culture of newsrooms." (6)

What are the main factors involved in the "breaking down" of journalism? The present transformation is largely a result of numerous factors simultaneously in existence: declining advertising income, financial crises of traditional news business models, the presence and influence of online media, the public's growing involvement in the gathering and sharing of news, and the consequent changes in traditional routines of news flow. Commenting on the findings of a 2008 survey of journalists also conducted by the *The Pew Research Center's Project for Excellence in Journalism*, among American professionals in 2008, Rosenstiel and Mitchell (2008) wrote: "Journalists have become markedly more pessimistic about the future of their profession. But their concerns are taking a distinctly new turn. Rather than worrying as much about quality, they are now focused on economic survival. And in that new focus, we see signs of new openness to change" (20).

In fact, as the results of the survey show, "the financial crisis facing news organizations is so grave that it managed to overshadow concerns about the quality of news coverage, the loss of credibility and other problems that have been very much on the minds of journalists over the past decade" (Pew, 2008, 1). In the report, 55% of American journalists cited financial or economic concerns as the most important problem facing journalism, up from 30% in 2004. In the search for causes, most journalists blame online media. However, as the survey states, journalists "make clear distinctions between the Internet's impact on the news business, which they view with alarm, and the ways that the web has transformed journalism, many of which the journalists view quite positively" (Pew, 2008, 2).

Indeed some see technology as a boon for helping journalists overcome the sense of isolation and distance from their audiences that was a key element of the so-called credibility crisis in journalism a decade ago. According to the Pew survey, American journalists expressed generally positive opinions about technology-driven changes in news production and delivery, namely giving the reader or viewer more news choices and transparency and the ability to weigh in on the quality of news. However, they continued to worry that time pressures may lower the quality of reporting, and of the increasing difficulty to determine the credibility of online news coverage. Interestingly, in any case, a good majority of journalists (66%) see themselves in the traditional role of gatekeepers, and think that it benefits society for them to continue to do.

A second factor accounting for the present transformation of journalism stems from the audience. The notion of a "public" has shaped journalistic values for centuries. At first addressed as consumers of popular newspapers, readers quickly turned into *citizens*, willing to be informed as part of their democratic responsibilities. This idea of the public influenced the notions of journalists as watchdogs and professional antagonists of power, protected by the values of objectivity, balance, accuracy and a strict separation between news and commentary. Different professional models, however, developed inside these boundaries: *neutral journalism*–strongly attached to the values of objectivity; *interpretative journalism*–interested in giving context and analysis to the news; *advocacy journalism*–concerned with making sure different voices have access to the public sphere; *watchdog journalism*–the investigative role directed to the control of political power; *new journalism*–[specially since the 60's, but with strong predecessors since the beginning of the 20th century], mainly dedicated to developing stories generally neglected by mainstream media in a literary language and structure; and the more recent *public journalism*– concerned with the participation of the community in the building of media agendas (Ortega and Humanes, 2000, 118). In all cases, journalists shared the image of a more or less homogeneous audience as primary target of their work, which was generally conceived as a public service towards those citizens.

In recent decades, however, the notion of a public has shifted once again. Not only did the "public" start to show diversity, atomization, inattention to public matters and cynicism towards politics and journalism, but technology also cultivated a new type of audience: engaged, opinionated, active, and collaborative. These new audiences—no longer simply readers—are changing the way journalists see themselves. Those who see the hypermedia landscape as a synonym for chaos and a random proliferation of data tend to claim that journalists' ability to tell truth from bias and to act as watchdogs has been compromised. But those who happily embrace the new attitude of active audiences point out a new age for journalism, where professional reporters sacrifice part of their power as mediators to collaborate with their audiences in a new public sphere for news.

Journalists, however, continue to struggle to adapt to the new realities of their profession. When asked about the state of journalism, 62% of American professionals say it is "going in the wrong direction", up from 51% in 2004 (Pew, 2008). Journalists surveyed believe that news media have cut back too much on the scope of coverage, pay too little attention to complex issues and have blurred the boundaries between reporting and commentary.

While these trends may fit journalists' perceptions in newsrooms worldwide, access to technology for both professionals and audiences varies dramatically in different geographies. According to the report Measuring the Information Society, by the *International Telecommunication Union* (2010), in 2009 an estimated 26% of the world's population (1.7 billion people) had access to Internet. But in developed countries the percentage remained much higher than in the developing world, where, four out of five people were still excluded from the web. Also, while Internet penetration in developed countries reached 64% at the end of 2009, in developing countries it reached only 18% (and only 14% if China is excluded). In other words, the digital divide continues to exert influence in the development of a cohesive global portrait for 21st century journalism.

Other differences may be found in the media industry as well. As a 2010 PricewaterhouseCoopers report on media business has shown, although the global newspaper market fell by 11.4% in 2009, and further decreases are expected in 2010 and 2011, such numbers can be misleading. Rising circulations in Latin America and Asia Pacific are offset declines in paid circulation of newspapers in North America, Europe and the Middle East. According to the report, these regions' newspapers will face less competition from the Internet and will also benefit from the expected growth in the number of people 45 years of age and older, the primary newspaper reading demographic group (PricewaterhouseCoopers, 2010). These findings certainly question the idea that new media are dwarfing the role and importance of traditional media.

While the technological advancements of the 21st century have brought about increasing uncertainty in newsroom around the world, at the same time they have enabled new opportunities to redefine the ways in which journalists maintain their core purpose of informing an active and engaged public. As new pressures mount to secure finances and audiences, journalists are finding new avenues for collaboration, dialog, and relevance.

Three Issues Shaping Journalism in the 21st Century

The state of uncertainty for journalists is connected by a common purpose to continue to find ways to inform and engage citizens about germane issues that contribute to democratic discourse and civic participation. Three specific issues at the core of the journalism can help to explore some of the new realities that journalists face in the 21st century. Specifically, they are: *objectivity as a professional value, the journalist-public relationship* and *journalists' as social curators of reality.*

1. Objectivity as a Ritual of Practice

Objectivity has been at the core of professional journalism for decades, both as a practical and ethical goal. Since the dawn of the 20th century, neutrality and distance were accepted as the surest ways to present facts to the public in a credible manner, while avoiding accusations of bias or partisanship. The separation between the news pages and the editorial pages was widely adopted in the mainstream press, resting on several overarching principles: reporting can and should be separate from any special interests; facts should be separate from editorial; and the public must have neutral information to form their own opinions and make personal decisions. Objectivity thus shaped one of the most extended and long-lasting journalistic models: the "neutral journalist", who can avoid personal judgments and is able to report life as if he was not there to do so (Abril, 1997; Fuller, 2002).

As many have pointed out, in everyday journalistic practice, objectivity is helpful, namely as "an expressive and strategic ritual" (Chillón, 1999, 47), or "a form of persuasion" (Rosen, 2010, 1). In other words, it is present both in the rhetorical artifices and writing techniques of journalists. Wrote Rosen (2010), "journalists have kept objectivity more or less the same over the years: a system of signs meant to persuade us to accept an account of what happened because it appears to contain only what happened and not what the composer of the account feels about it" (2). At the same time, objectivity may serve as *"a strategic ritual of defense"* (Rodrigo Alsina, 1993, 169), helping to sustain the work and functions of the newsroom.

However, even a quick look at the way in which mass media function and the growing complexity of the issues journalism deals with today may support those asserting the impossibility–and even negativity–of struggling to be objective. Neutrality has even been contested as negative in regards to the public interest (Malkki, 1997; Restrepo, 2001; Fuller, 2002; Kapuscinski, 2003).

In today's digital age it is widely accepted that media in general and journalism in particular not only do construct the information they gather, but also shape and shift facts. There are various mediations on the way media create and frame information, such as the particular angle from which journalists stand to look at reality; the objectives, structures and routines of the newsrooms where they work; the technical and economic conditions of their work; and the conflicts of interests and power struggles that the social circulation of information implies (Meditsch, 2005). "While the rules of objectivity forbid journalists to make subjective judgments, their very same task demands it. Every fact is a cultural construction and can only be communicated through its place in a system of signs that is shared by journalists and readers" (Malkki, 1997).

While issues with objectivity have existed for some time, new media technologies have exposed them now more than ever before. The illusion of objectivity weakens when one considers the inevitable subject-driven decisions journalistic work demands, such as which facts get coverage, which sources are cited, what information is accepted to be published or not, and the images, videos, headlines and opinions that complete the coverage.

Thus, accepting the conditions of objectivity and making them apparent to the public in both traditional and digital spaces may be more positive than hiding in the shadows of transparency. Does this mean journalists should forget about honesty, accuracy and balance? Absolutely not. Original reporting and verification continue to be strict priorities for journalists. But, in times of increasing public scrutiny and a pervasive flow of information through digital and multimedia platforms, acknowledging the critical capacity of journalists may be beneficial to an active audience. As Rosen (2010) writes, "instead of two rigid poles—neutrality or ideology, news or opinion, reporter or blogger, adults or kids—I recommend a range of approaches that permit journalists to report what they know, say what they think, develop a point of view in interaction with events and bid for the trust of users who have many more sources available to them" (4).

Journalists must consistently maintain a delicate balance between investigation and distance. This back and forward movement continues to be the best way of building trust and credibility no matter which platform the information flows through. Objectivity is not safe: transparency asks for a different bid of trust, one that new audiences may be ready—even eager—to accept more willingly.

2. New Public? New Journalism?

Atomization and independence are the visible marks of today's audiences. Increasingly engaged citizens recognize the value of information for the public good. The goals for journalism have thus multiplied. Stated the Pew Center (2010), "reportorial journalism is getting smaller, but the commentary and discussion aspect of media, which adds analysis, passion and agenda shaping, is growing, in cable, radio, social media, blogs and elsewhere...argument rather than expanding information is the growing share of media people are exposed to today." The same report then goes on to notice one of the permanent values of traditional media:

> "The numbers still suggest that these new media are largely filled with debate dependent on the shrinking base of reporting that began in the old media. Our ongoing analysis of more than a million blogs and social media sites finds that 80% of the links are to US legacy media." (Pew, 2010)

In the same spirit, the "Digital Future Report", published annually by the Annenberg School for Communication and Journalism (USC), found "large percentages of users who express deep distrust in online information" (Center for the Digital Future, 2010). In other words, the audience is more diverse, independent, willing to receive and give opinion, less open to facts, and evermore insulting behind the anonymity the web allows. Nevertheless, journalistic abilities to report accurately, give context to events and unearth more facts remain useful and in demand. What has changed is that journalists must now co-exist with a growing cohort of information players, ranging from social media to self-interested non-journalistic voices, contributing to a rapidly growing news field. There are even some analysts foreseeing partnerships between mainstream and amateur media, although, "how traditional news organizations cope with such partnerships, the rules for what is acceptable and what is not, remain largely uncharted" (Pew, 2010). In the meantime, while more is learned about the economics of the web and online consumer behavior, it is

worth remembering that journalists have always been trained to strike a balance between what audiences want to know and what is important for them to
know. In a media atmosphere that seems more willing to achieve customer
satisfaction than civic engagement, journalism can claim that ability as its contribution to the hypermedia age.

3. From Gatekeepers to Social Constructors of Reality

Journalists have always seen themselves as mediators between public issues and
their audiences, "gatekeepers" with the power to check, interpret and deliver
what people need to know. However true this may sound today, such a pervasive self-image may be somewhat out of fashion or even naïve in the present
media environment.

The role of the media as social constructors of reality has largely been accepted in the academy and increasingly among more skeptical audiences. Media are now understood as key references in constructing the sense of the
broader social forces that affect people's everyday lives, providing orientation
for their beliefs and structuring their personal experiences. However, media
operate in a web of relationships and power struggles with other institutions
that also shape content, namely government, advertisers, public relations professionals, influential news sources, interests groups and other media organizations (Reese, 2001). This process is not usually exposed. On the contrary,
ideological orientations and editorial guidelines are implicitly reproduced in
newsrooms mainly through the routines of everyday work.

While scholars have analyzed news production for well over a century, the
idea of journalists making news instead of merely reporting happenings may
still be difficult to accept within the profession. To a large extent, traditional
journalistic training continues to emphasize the role of journalists as professional antagonists to political power, the rituals of mistrust and the commitment to truth, while both trainers and trainees find themselves surrounded by
a media environment that rarely reflects such ideas. If governments, political
parties and public bodies have always struggled to establish their own agendas,
new media have enabled them to reinforce their power to do so. Technology
seems to be shifting power to newsmakers, mainly through their ability to control the initial account of events. As the Pew Center (2010) states:

> ...shrinking newsrooms are asking their remaining ranks to produce firsthand ac-
> counts more quickly and feed multiple platforms. This is focusing more time on dis-
> seminating information and somewhat less on gathering it, making news people more

reactive and less proactive (....) What is squeezed is the supplemental reporting that
would unearth more facts and context about events. (7)

Social constructions of reality are not only in the hands of journalists, but
at the core of a political struggle involving different social forces that paradoxically takes place mainly in the media arena. Inside media organizations—that
are of course an active part of the struggle, on behalf of their own economic
and political interests—journalists need not only to acknowledge their role as
constructors of daily reality, but also to recognize their part in a battle being
fought in their own backyard. While journalists still find the idea of constructing reality rather than merely reporting the news difficult to support openly,
they now find themselves obliged to share that social power with other players,
many of who do not have the notion of public service at the core of their
agendas.

The disputes around objectivity as a practical and ethical goal, the am-
biguous relationship between journalists and the public, and the increasingly
contested role of gatekeepers of information have together renewed the relevance of journalism in the context of the pervasive technological influences on
society.

Conclusion: A Profession Reshaped: News Literacy and 21st Century Journalism

Although the gloomy forecasts for journalism are normally reinforced by a
seemingly reliable set of statistics and cases studies, it would be a mistake to
listen only to those voices that continually predict the downfall of the profession. In today's newsrooms—physical and virtual—the scenario looks somewhat
different. The following conclusions (in point-counterpoint structure) represent a framework of new realities for journalism in the 21st century.

Point: Journalists are no longer the only voices...

Over a relatively short period of time, the world of information has expanded
exponentially. Not only in terms of the increasing amount of data, opinion,
argument, commentary, images and words that endlessly circulate, but also in
the number of citizens around the world are now participating, sharing and
collaborating across media platforms. At the same time, the sheer number of
media outlets available has multiplied, which has provided journalists more
avenues for dissemination and commentary. Journalists should anticipate

smaller audiences, with precise interests and needs, and even names and faces, instead of a large crowd of unrecognizable readers, always willing to listen with no immediate capacity to respond.

Counterpoint...but not everyone is on the web

Due to social inequalities, lack of education or disinterest in public matters, life for many continues offline. Journalists remain responsible for supplying all citizens with information, making their voices and needs heard and accounted for, and reflecting their interests as well. An important task for journalists to-day is to indentify specific audiences rather than simply provide information.

Point: Journalists no longer solely decide what is news, or report it first...

How news is presented, disseminated and received today has significantly changed. Being able to explain, give context, and translate information for an audience has become increasingly important, while the impetus to "break" news has subsided.

While many voices with varying intentions struggle to gain space and time in the media, journalists still have the responsibility to provide credible information and honest commentary, and to give space to relevant voices. In this context, journalists in the 21st century must understand the social media tools and converged platforms for which they can cultivate active and dynamic relationships with their readers, to provide consistent, relevant, accurate, and personal information for certain communities. In this way, journalists are no longer seen as news breakers, but rather as news curators.

Counterpoint...but journalists still know how to act as watchdogs

The democratic duty of acting as watchdogs has not ceased for journalists. In fact, it has become more critical as political powers have also learned to use new media to their advantage. Society still expects journalists to be the bridge between government and the public. In that sense, most traditional journalistic values still matter. Objectivity should be openly replaced by honesty and sensibility. Far from diminishing, post-objective realities for journalists have achieved critical relevance in digital spaces, where the illusions of transparency and direct interaction with sources of information seldom make professional journalism appear useless.

Point: Journalists should think of news as borderless...

Even if we recognize that inequalities are excluding entire populations from new sources for news, information circulates today with no regard for boundaries. Journalists must understand their reach in all aspects of their work: not only because what is happening close to one's newsroom might have an immediate impact in local communities, but also because other reporters in different places may have access to valuable information to share.

Counterpoint...but journalists cannot rule out local perspectives and needs

In their struggle for conquering new audiences, mainstream media have lately found a strategy: telling human stories or finding the human angle in hard news. However over-used this strategy is, there remains a certain value to it: as politics, economy and international matters become more complex, people tend to be interested in what happens to other people. In Latin America in particular, the re-emergence of feature articles in magazines and some newspapers ("crónicas"), with extensive field reporting, a deep interest in picturing neglected populations and experiences, and literary writing, has given voice to those silent communities in the increasingly global world.

Point: The Internet is an attractive new arena for journalists...

As mainstream newsrooms shrink and working allowances for journalists lessen, the web has become an attractive space for "traditional" journalists. Journalists today create their own identities online, where they can deal with issues and angles commonly ignored by the mainstream media. Having one's own magazine, radio show or documentary program has never appeared so easy. Lately, even journalists still in traditional newsrooms have found new spaces for reporting online.

Counterpoint...but journalists must continue to make a difference in cyberspace

Enthusiastic approach to the web has not always been accompanied by reflection on what to offer the public through the Internet. The amazement with which many media companies and journalists welcome the web seldom overlooks the extraordinary amount of reflection, creativity, experimentation, and mistakes that traditional media have gone through in their transition online. Thus, reshaping journalism is not as much about adapting to a digital landscape but about rethinking practices and representations that have been the status quo for some time. That journalists must share the decisions about what's news with an increasing number of civic voices should not rule out journalism's ongoing responsibilities as watchdogs and to alert society of key phenomena that might remain otherwise unnoticed. With or without technology, inside or outside the Internet, journalism continues to be mainly about telling stories that are critical to understand the world we live in.

REFERENCES

Abril, G. (1997). *Teoría General de la Información*. Madrid: Cátedra.

Center for the Digital Future. (2010). *Digital Future Report*. USC Annenberg School. Retrieved October 15, 2010 from: *http://www.digitalcenter.org/pages/current_report.asp?intGlobalId=19*

Chillon, A. (1999). *Literatura y Periodismo*. Barcelona: Universidad Autónoma de Barcelona.

Fuller, J. (2002). *Valores periodísticos. Ideas para una Era de la Información*. Miami: Sociedad Interamericana de Prensa (SIP).

International Telecommunications Union. (2010). *Measuring the Information Society*. Geneva: ITU. Retrieved 15 October 2010 from: *http://www.itu.int/ITU-D/ict/publications/idi/2010/index.html*

Kapuscinski, R. (2003). *Los Cinco Sentidos del Periodista*. México: Fondo de Cultura Económica.

Malkki, L. (1997). "News and Culture: Transitory Phenomena and the Fieldwork Tradition". In Gupta, A. and Ferguson, J. *Anthropological Locations*. Berkeley, CA: University of California Press.

Meditsch, E. (2005). "Journalism as a Form of Knowledge." in *Brazilian Journalism Research*, 1(2), 121-136.

Ortega, F. and Humanes, M.L. (2000). *Algo más que Periodistas. Sociología de una Profesión*. Barcelona: Ariel.

Pew Research Center For The People and The Press (2008). *The Web: Alarming, Appealing and a Challenge to Journalistic Values*. Retrieved 10 October 2010 from: *www.journalism.org*

——. (2010). *State of the News Media 2010*. Retrieved 10 October 2010 from: *www.journalism.org*

Pricewaterhouse Coopers. (2010). "Global Entertainment and Media Outlook 2010-1014." PricewaterhouseCoopers Report.

Reese, S. (2001). "Understanding the Global Journalist: A Hierarchy-Of-Influences Approach," *Journalism Studies*, 2(2), 173-187.

Restrepo, J.D. (2001). "La Objetividad Periodística: Utopía y Realidad". *Chasqui* 74. Quito, Ecuador: CIESPAL.

Rodrigo Alsina, M. (1993). *La Construcción de la Noticia*. Barcelona: Paidós.

Rosen, J. (2001). *What Are Journalists For?* New Haven: Yale University Press.

Rosen, J. (2010). "Objectivity as a Form of Persuasion." Retrieved from Pressthink on 10 October 2010: *www.journalism.nyu.edu/pubzone/weblogs/pressthink/2010/07/07/obj_persuasion.html*

Rosenstiel, T. & Mitchell, A. (2008). "2008 Journalist Survey: A Commentary on the Findings". Retrieved October 10, 2010 from: *www.journalism.org*

Part Two

Pedagogical Models for News
Literacy Education

Chapter 5 -
Creating Shared Dialog through Case Study Exploration: The Global Media Literacy Learning Module

CONSTANZA MUJICA
Pontificia Universidad Católica, Chile

Introduction: Media Literacy Pedagogical Objectives: A Broad Definition

As media become increasingly central facilitators for our local, national and global communities, how educational institutions create media learning experiences becomes vital to civic democracy in the 21st century. This not only includes how educators teach students to analyze and interpret messages, but also how they teach about media as the key conduit for understanding cultures around the world. The common term to describe these educational efforts is media literacy. This chapter explores media literacy approaches to teaching about global news and media that are predicated on creating shared dialog and cross-cultural tolerance.

Most media education scholars coincide in broadly defining media literacy as the education of children and adults in the critical appropriation of the content, language, production context, and audience effects of the media. Hobbs & Frost, for example, summarize prior definitions and understand media literacy as "the ability to access, analyze, evaluate and communicate messages in a wide variety of forms' (Aufderheide & Firestone, 1993), emphasizing the skills of analysis, evaluation, and creation of messages that make use of language, moving images, music, sound effects, and other techniques (Masterman, 1985; Messaris, 1994)" (Hobbs & Frost, 2003, 334). Media literacy, says Buckingham (2003), is a critical literacy that:

> involves analysis, evaluation, and critical reflection," that is possible only through the "acquisition of a 'metalanguage'—that is, a means of describing the forms and structures of different modes of communication; and it involves a broader understanding of the social, economic and institutional contexts of communication, and how these affect people's experiences and practices (Luke, 2000). Media literacy certainly includes the ability to use and interpret media; but it also involves a much broader analytical understanding. (38)

Buckingham (2003), like Tisdell (2008) and others described by Mihailidis (2008) and Oxstrand (2009), adds to the critical analysis of media messages the production of media content as a key aspect of media literacy. "...Media literacy necessarily involves 'reading' and 'writing' media. It enables young people to interpret and make informed judgments as consumers of media, but it also enables them to become producers of media in their own right" (Buckingham, 2003, 4). Tisdell asserts that "nearly all critical media literacy scholars focus to some extent on people's appropriation of media messages, either as active consumers of media, when one specifically chooses to play a video game (Gee, 2004) or watch a movie, or the more passive consumption that happens by unavoidable interactions with the media just by living in the world, such as in passing by billboards or having the radio on in the car or hearing the television someone else in the house is watching" (Tisdell, 2008).

Buckingham adds nuance to the sometimes generalization-prone concept of media literacy when he speaks of multiliteracies. Media literacy is not comprised of only one type of media appropriation technique or understood within a finite amount of content. Rather, Buckingham (2003) believes that there are multiple media languages and multiple social contexts and appropriations, because "different social groups define, acquire and use literacy in very different ways; and that the consequences of literacy depend upon the social contexts and social purposes for which it is used" (p. 38).

When speaking of the construction of media literacy experiences, Buckingham chooses to speak loosely of four key concepts: production, language, representation, and audience. He understands them not as a blueprint for media concepts or contents, but as non-hierarchical and interdependent. Each concept is a point of entry into the analysis and creation of media contents that invoke all others.

For Buckingham, the study of production is based on the understanding that media are consciously manufactured, usually for commercial profit, increasingly on a global scale, and, thus involve economic and political interests. For him, production involves the critical analysis of technologies, professional practices, industry, connections between media, regulation, circulation and distribution, access and participation. Additionally, the concept of language considers the study of meanings, language conventions, codes, genres, choices involved in the construction of any media message, and the combinations of codes and choices. Representation, according to Buckingham, considers issues of realism, objectivity, and truth, presence and absence of groups of people, stereotyping, interpretations, and influences. Finally, audience considers targeting, address, circulation, uses and appropriation, making sense, pleasures, and social differences.

Duran et al. (2008) also consider a set of general concepts and issues that, "from a holistic, critical, contextual perspective, it may be argued that a media literate person understands" (51). They list six such general questions: what types of persuasive messages are found in media; why media messages look, sound, and read the way they do; who creates and benefits from these messages; when are we affected by media; where can we find alternative media; and how can we actively work to change the media system (Duran et al., 2008). Tisdell (2008), quoting Yosso (2002) further summarizes the assumptions behind the definition of these general areas:

> "the media are controlled and driven by money; (b) media images are constructions—both of directors, actors, and other media makers; (c) media makers bring their own experience with them in their construction of characters, including their perceptions of race, gender, class, and so on; (d) consumers of media construct their own meaning of media portrayals in light of their own background experience and gender, race, class, or sexual orientation; (e) unlike print media, entertainment media, such as movies and television, are a combination of moving visuals, sounds, and words that combine in facilitating meaning; and (f) it is possible to acquire multiple literacies in becoming media literate." (2008)

Mihailidis (2008a, 2009) also lists a foundation of critical skills: comprehension, analysis, evaluation, and production of media. However, based on focus groups with college students that had participated in media literacy classes, he argues that at least in post-secondary environments the exclusive focus in these kinds of skills might lead to a cynical point of view about the media. Participants showed negative perceptions about the role of the media in democratic societies, and viewed being media literate as a way to avoid falling in the media's "fake and manipulated" images of the world. The author asserts that "the overwhelming evidence points to a media literacy experience that effectively taught the students skills to critically view media, but not how such critical viewing should be couched in media's larger civic roles and responsibilities" (Mihailidis, 2009). Mihailidis worries that this perception crosses the line between skepticism and cynicism:

> "The promotion of healthy skepticism—consistent inquiry concerning how media portrays cultural, social, political and economic issues, coupled with a general understanding of the media's role in civil and democratic society—is at the center of media literate learning outcomes. Media literate individuals, it is often purported, should be

open to different ideas, demand evidence for certain claims, and approach information with a keen sense of interest, independence, and awareness. In this sense, if media literacy is to enable a healthy skepticism towards media and information, it must
not only teach the skills of critical analysis, but also teach how those skills are pur-
posed around modes of general inquiry. This centers on making media literacy *pur-
posive* by highlighting the connections between media analysis and a nuanced
understanding of media's role in community, civic life, and democratic society." (7)

Thus, he suggests a model of media literacy education "that transfers the focus
from skill attainment to qualitative learning outcomes. Media literate students
should understand the social influences of media, reflect on the complex functions of the media, and be aware of the civic necessity of a media system" (Mihailidis, 2009).

Case Studies as a Teaching Method: Addressing Media Literacy Skills and Learning Outcomes.

The brief overview above covers just a few of a myriad of papers, conference
proceedings, reports, articles and books trying to explain media literacy and
the relevance of educating students and citizens to critically view media. Literature on teaching approaches to achieve these outcomes is even less common and clear.

Buckingham (2003) suggests numerous classroom activities to address each of
the key concepts he describes: production, language, representation, and audience. To deal with production issues he proposes research-based tasks on specific
production issues, and the application of general issues to the student's own media creations. Language can be addressed through close observation and analysis,
the application of semiotic concepts and research, the comparison of particular
texts, and debate arising from student media creations. He suggests that representation can be tackled through classroom debate on issues of realism, truthfulness,
presence and absence of individuals, social groups and points of view, bias and
objectivity, stereotyping, interpretations, and influences. Finally, audience can be
dealt with through classroom dialog, self-reflection and firsthand research.

Buckingham then systematizes and describes six classroom strategies for a
media literate classroom: textual analysis, contextual analysis, translations,
simulations, production, and case studies—that taken together reflect a comprehensive framework for a holistic media literacy educational experience.

a) **Textual analysis** involves the detailed critical study of single texts. Stu-
    dents are encouraged to provide evidence for their judgments about spe-

cific media texts through attention to detail and rigorous questioning; through making the familiar strange.

b) **Contextual analysis** relates specifically to texts with the political, social, and economic context in which they were produced, thus linking the analysis of language and representation with the study of production and audience. In this kind of work, students can revise the ways in which media products are targeted to certain social groups or genders, who produces them, how they are received by audiences and how they relate to others.

c) **Translation** refers to the interest in the content changes when treated in diverse media languages or genres. This approach goes from the purely analytical—investigating the treatment of an issue in two different media— to the practical, the production of a new media text dealing with an event that other media have already covered.

d) **Simulations** are modes of role-playing in which students are put in the position of media producers in a fictional way. They usually include exercises that entice participants to imagine they are editors, journalists, directors or producers.

e) **Production** involves a hands-on teaching perspective. Students are asked to create their own media texts as a way to understand, among other elements, conditions of production, and the possibilities and limitations of media languages.

f) In the creation and consideration of **case studies** students are expected to do in depth research on a certain media topic or issue as it develops in a specific situation of their choice, where all stages in the preparation and delivery of the case study are led by students and guided by teachers.

Thus, Buckingham sees the case study approach as not only as a learning opportunity for its final users, but also for the people who create them. The identification of relevant media issues and events, the packaging of them for the use of other students, the research required to elaborate educationally rich stories and exercises, and the debate with other students and faculty to eliminate prejudice and bias from the case study creator's point of view are all in and of themselves part of a rich learning process. Through the creation of classroom tools for others, the authors learn their own limitations and biases, while actively, critically and passionately engaging with media issues.

For Buckingham case studies are especially useful to link all of the key concepts he describes: they can analyze the production and context of one particular media text, compare the treatment of one issue by different media languages, and discuss the reception of it by the audience. For this reason case studies must deal with a great variety of data and sources, and require students to have research abilities, specifically gathering, evaluating and repurposing information.

Although textual and contextual analysis, and production will be described in detail in the subsequent chapters as separate teaching methods, the global media literacy case study learning modules created by students at the Salzburg Academy on Media and Global Change apply most of these methods in three complementary parts: the narration of a case in a story, the proposal of a classroom exercise that addresses the media issue that arises from the story, and the discussion of the general media concepts that both story and exercise propose through previously defined analysis criteria.

Global Media Literacy Case Study Learning Modules

At the Salzburg Academy on Media & Global Change,[1] over 50 students and a dozen faculty gather annually in Salzburg to work in international teams and across disciplines on a dynamic web-based curriculum titled "Global Media Literacy." Students, representing over 15 nationalities on five continents each year, build web-based lesson modules on how global media cover key issues and work to provide portrayals of cultures beyond physical borders.

The module topics are chosen by the Academy participants themselves with guidance from faculty, and consist of various parts—story, exploration, reflection—which together aim to provide an engaged, critical and diverse look into how media influence civic societies around the world. These curricular products are premised on building global understandings of culture, ideology, and diversity through media literacy.

Part 1: The Story

The first part of the case study model developed by the Salzburg Academy is the short narration of an individual situation or event that is selected as an example of a broader media issue. The major advantage in allowing the creator to choose the topic and angle of research is the sense of pleasure brought by investigating the student's own interests, which can "stimulate critical reflection on assumptions, which is part of transformative learning" (Tisdell, 2008). Tisdell's research shows that student's passion for a topic or event encourages them to enthusiasti-

---

1 For more information on the program, see www.salzburg.umd.edu

cally go deeper into the media issue under study, and to find new facts and points of view that might question their own preconceptions.

But, as both Tisdell and Buckingham recognize, this sense of pleasure may lead precisely to an insistence in the student's preconceptions and the loss of the broader issue behind the story (Tisdell, 2008; Buckingham, 2003). Wrote Buckingham (2003):

> It is important that students recognize that a case study is an example –it is, precisely, a case study of broader issues or tendencies. Media education is not a license for students simply to accumulate vast amounts of information about their media enthusiasm. They need to be encouraged to recognize the broader issues that are at stake in them. (p. 77)

To avoid this risk, the selection of topics and issues for the case studies can be oriented by suggesting topics or issues that are explicitly related to the critical skills and learning outcomes described by Mihailidis (2008b) and the Salzburg Academy on Media and Global Change. Students can, for example, select topics linked to the identification of what "news" is and how media, as well as other actors, decide what information matters—such as citizen journalism, covering democracy, global issues, local coverage, or news credibility—to the monitoring, analysis, and comparison of media coverage of people and events, such as the study of the use of graphic images, the coverage of institutions, or media ownership; and to the understanding of media's role in shaping global issues, such as the coverage of terrorism, environmental issues or the representation of them in entertainment media.

To highlight the connections between media literacy and civil society students can study issues related to the defense of the media's role in the "oversight of good government, corporate accountability and economic development," such as media at risk, government watchdogs, and free press during crisis; the promotion of "civil society by themselves becoming a responsible part of the communication chain," such as citizen journalism, alternative media, and media and education; and the motivation of media to "better cover news by communicating to media their expectations for accuracy, fairness and transparency," such as media and science, entertainment media and social responsibility, and social marketing (Salzburg Academy on Media and Global Change, 2008).

These issues can be collectively defined as the so-what factor: the broader media issue that the story is an example of and the element in the story that makes it useful and appealing to a broader audience beyond the student's personal interest. The so-what factor should be explicitly stated in the beginning of the story and all details that are included in it need to be clearly related to this broader issue. It is not enough to choose an exciting topic—the death of young people in a fire or the protests of high school students for better education standards—rather, it is essential that the topic is linked to an equally interesting media problem. For example, the use of graphic images in the media or the portrayal of protesters as criminals in traditional media and as future leaders in the blogosphere are topics in which a worked example can be extrapolated to relate to a larger issue.

Students also must be encouraged to think globally about their case study and media issue. As has been discussed in prior chapters, the emergence of digital media and mobile technologies has made possible an unprecedented flow of information globally. Information and opinions expressed by someone in a small town in any country in the world can be shared, commented and re-published worldwide. This means that very local issues and personal experiences with media can also serve as global media examples given that the story is explicitly linked to broader media themes and values. For example, a student of the Salzburg Academy on Media and Global Change included in her case study module the exchange of emails between two young women, one Lebanese and the other Israeli, during the 2006 war between both countries. The personal correspondence between them was used as a way to show how these women "wanted to understand the war from each other instead of through the mainstream media. They felt that the available media sources in a given country during conflict only offer a single side, rather than a balanced view, of an event" (Hobollah & Jing, 2008).

The presentation of such experiences can spark interesting debate that can encourage the anchoring of the story to a global so-what factor. Such discussions allow students to go beyond their own sense of place and understanding, and permit a deeper analysis "on critical media literacy and diversity issues" (Tisdell, 2008) and "encourage a more distanced, reflective approach" (Buckingham, 2003). Other students, coming from different ethnic, social or cultural backgrounds can posit doubts about the relevance of the story to the general media issue that the author of the case study is proposing, suggest other issues linked to this main issue, and ask for the inclusion of new details relevant to the understanding of the story.

Another key issue to be considered in the construction of a case study story is which additional information and multimedia resources to use. The global reach of many current media events, and the advantages of digital platforms for resources increases the temptation to construct case studies full of images, videos, and references to web sites. Students should ponder how ubiquitous access to computers and broadband is in other countries and, thus, if other students will be able to view and use that material. In that context, teachers and students must discuss if those resources are relevant to the story, and, if they are indispensible, what alternatives can be offered so the lesson module does not suffer from short-sightedness or a lack of adaptability.

Integrated resources and multimedia can serve as the bridge between the specific example under study and how it can relate to similar issues present in other parts of the world. This is shown in the following examples:

The Case Study Story: A Commented Example

Example 1[2]:

Case Study: The Power of Citizen Journalism Demonstrates the Need for Standards

By: Michelle Leibowitz

The field of citizen journalism has exploded in recent years, due in large part to technological advances. There are a variety of forums available for citizens to contribute news and for professional journalists to draw news from. The problem with this seemingly efficient and enhanced news system lies in how much credibility to afford the citizens' contributions. Media outlets worldwide have a variety of ways they can present citizen journalism. They must constantly keep in mind how their presentation and framing of information affects how consumers of news interpret that information.

On October 3, 2008, news outlets in the United States showed just how important the presentation of citizen journalism is. A user on *CNN's* citizen journalism platform, *iReport*, posted that Steve Jobs, founder and CEO of Apple, "was rushed to the ER after suffering a major heart attack"

A variety of news sources picked up the story. Some, such as *Gawker* (an online breaking-news site), reported it with a healthy degree of skepticism, but other news outlets presented the information as credible, causing a drop in Apple's stock price according to *CNET.com*.

> **The so-what factor:**
> The amount and variety of citizen media make it difficult to determine standards to assess the credibility of citizen media and citizen contributions on traditional media.

> **Details linked to the so-what factor:**
> A detailed chronology of the propagation of the news from a citizen fueled traditional news site to finance specialized media.

The post appeared on *iReport* early in the morning Eastern time (USA). By 10a.m. Eastern time/ 7a.m. Pacific time, according to *CNET.com*, the stock had dropped. Because of how early the information was posted, Apple's California offices were not yet open to declare the report false. *Business Insider* reported that the stock did later that day recover from its 5.4% drop after the information was revealed to be untrue.

If the Jobs' rumor had been posted to a less prestigious message board somewhere else in cyberspace, it may not have been taken as seriously as it was. Because *iReport* is associated with *CNN*, and because of the context in which the user-generated content was presented, other news outlets afforded the post the credibility that comes with being associated with *CNN*. When news outlets such as *CNET.com* (a popular and respected American technology site) and the *Silicon Valley Insider* (a business outlet that covers the Silicon Valley area in California) passed on the information, their weight and reputation added to the perceived truthfulness of the *iReport* post.

CNN's policy regarding *iReport* stated at the time: "we've launched an independent world where you, the iReport.com community, tell the stories we're not used to seeing. And the most compelling, important, and urgent ones may get seen on CNN." This "may get seen" wording implied that *CNN* would check the accuracy of statements that appeared on *iReport* before allowing them to be posted. Although *CNN's* policy at the time did also state "the views and content on this site are solely those of the iReport.com contributors. CNN makes no guarantees about the content or the coverage on iReport.com!" the earlier statement still suggested at least some oversight of published content.

Explain context to make it comprehensible to a broader and potentially global audience. Publishing context and considerations, like the association to a well known traditional journalistic medium, that increased the original publication's credibility and favored its reproduction in other media.

Once Apple's representatives learned of the *iReport* statement, they promptly denied the report and publically announced that Steve Jobs had not been hospitalized. *CNN* then called the posting "fraudulent" and removed it, but belatedly. Sources such as the *Silicon Valley Insider* posted Apple's denial a full 20 minutes before *CNN* took down the *iReport* item~a near eternity for the stock market.

Business Insider referred to this case as the "first significant test" for citizen journalism. The *iReport* story and its consequences gave rise to such issues as:

- Should citizen journalism be treated the same as journalism reported by professionals?
- Citizen journalists do not have the training and experience of professional journalist, so what standards should they be held to?
- How quickly should the public believe the content of citizen journalism?
- How skeptical should news consumers be about citizen journalism? Should professional reporters be even more skeptical?
- How much skepticism is too much?

Other media issues that arise from the original event.
This case can serve to discuss not only the difficulties to assess information produced by citizens, but it can also spark discussion about other matters like media credibility, journalism standards for citizen media or the value and risks of skepticism.

The lesson is that context matters. The circumstances in which citizen journalism is presented are vastly important. They influence whether information is viewed as credible, and how much benefit of the doubt information gets. Get it right, and the world reaps the benefits of infinitely more observations and opinions on an expanded range of issues. Get it wrong, and news consumers worldwide lose so much insight and information.

In a perfect world, some, citizen journalism should be held to the same standard as professional news organizations. Yet for most practitioners, citizen journalism is more of a hobby than a formalized way to contribute to news. If citizen journalism cannot realistically meet professional standards of transparency, then what? It would be incredibly unfortunate to lose out on all the information and observations citizen journalism provides.

If the information cannot be presented as confirmed truth, at the very least it could be presented as food for thought, to be taken with a grain of salt. With careful consideration to how items are presented, citizen journalism can become a great venue for broadening the scope of news.

Part 2: The Explorative Exercise

Following the worked example of the case study module, students are asked to create exercises that place the broader media issue into active and dynamic explorations.

- After engaging with an issue in depth, students are asked to undertake an active exploration of the issue, followed by a discussion framework that helps reflect on the exercise. Exercises can be in the mold of role-playing, research, production, or comparative analysis. The benefits of this method are active engagement, inquiry and extrapolation of a story into wider contexts. As Tisdell points out "such an activity encourages learners to engage with the text (movie or television show) and often with others in such a way as to facilitate their own critical media literacy development" (Tisdell, 2008). Thinking about others educational needs forces students to "engage with the text more deeply....and further develop their own critical media literacy awareness.

- Many others also reported that having to develop such an activity forced them to get underneath what was obvious in a film or television show, thus clearly leading to their greater critical media literacy" (Tisdell, 2008).

- An exercise is more than just questions. Students must be asked to DO something, to engage practically with the media issue. Questions can be used to guide the activity or, at the end of the class, to sum up the conclusions each student arrived to.

Students can be asked to:

- Conduct simulations, for example to put him- or herself in a news editor's shoes through role-playing.

Example 2:

Case Study: Bridging Religious Divides~The 2006 War between Israel & Lebanon
By: Muneira Hoballah and Su Jing

Exercise:

Class discussion/Writing Assignment: You are a research assistant for a documentary filmmaker. You need to put together a short, 5-minute video package on a war in your region. The goal is to be fair and balanced and to foster mutual understanding between two sides of the conflict.

1. On the air, journalists often have a very short amount of time to get their story across. The filmmaker requires that you use these five minutes to illustrate who is involved, why they are involved, and what the outcome was.

2. You are to show the degree of devastation on both sides and appropriate the right amount of time to each side.

3. You will need to consider the following during the production process:
   - Who is your audience? What do you want them to know? Does the content or production of the film change depending on your audience?
   - Does it matter who was harmed more? Do you devote more time to the side that was more greatly affected? To the winner? The loser? Does it matter?
   - What images do you show for each side? Will you show graphic images?
   - What information is relevant to getting the story across? Will you focus on politics or religion? On history? On the violence?

Students are asked to address issues of fairness and balance in war and conflict reporting through the planning of a documentary.
They are expected to view the issue from the point of view of a researcher for a journalistic project. Questions are added to orient the generation of a project to ensure that this point of view is followed. Finally, these same questions can be used as a guide for class discussion.

- Discover and analyze textually and contextually similar cases on a local scale.

Example 3:

Case study: Eco-Journalism & Cleaning up Sumidero's Canyon
By: Andres Castillo and Walyce Almeida

Exercise
Writing Assignment: You want to see your local news outlet cover a social issue that has been overlooked by the media. Write a proposal asking your local newspaper, radio station or television station to cover the issue. Make sure you address the following questions:

1. What is the main problem that you want the media to focus on?
2. Why should the news outlet care about this issue? Why should its audience?
3. How can the story be told? How can it be made engaging?
4. Who are the sources for the story? Who are the voices who should be heard
5. What kind of background information is needed? How much of that is already known? How much additional research is necessary?
6. What is the visual or the aural side to this story—what can be photographed, recorded, videotaped?

> In this case, students are addressed as active consumers of media; they don't have to role-play. They are asked to take action on a local environmental situation by sending a letter to local media. Questions are added to center their latter discussion around the media elements of the event, and not only its environmental characteristics:

> why should media care about the issue—hence discussing newsworthiness elements of the story—sourcing, research and production of the news story.

• Research and compare different media texts.

Example 4:

Case study: The Cloning Fraud: How can media and science live together?

By: Mónica Uriarte

Exercises:

Divide the class in groups of three people. Assign each group to look at three different types of news media (radio, newspaper and TV). Have each group each day keep a chart of:

- The number of science/health and medicine stories and
- The number of sources each media used to cover each story in each news outlet.
- List the kind of sources in each story (politician, scientist, industry expert, etc.)

Here students are asked to follow a topic–science in this case—in diverse media outlets.
They are asked to evaluate, chart and compare the amount of coverage, the number and types of sources used.
Finally they must send their results and comments to the media they analyzed as a way to improve coverage.

Have the students keep a chart for a week. After the week has passed, allow students to share their findings. Discuss the following questions: What is the number of stories each media covered? How much time did each media spend in each story? How many sources each media used for each story? What differences did you find between science/health and medicine coverage with other topics such as politics, sports and entertainment?

Individually, write a letter to the station manager or editor of the media you found to be most "newsworthy" and explain what you appreciated most about their news coverage.

These types of explorations allow students to appropriate global or abstract ideas, and to adjust the distant to a culturally or socially specific context. This means, as was the case in the creation of the story, very close consideration to which online resources must be included and to the provision of alternatives that can be adaptable to a variety of technological, social and cultural contexts.

Part 3: The Analysis Criteria

Lastly, the global media literacy case study learning module format suggests new questions and considerations that arise from the general media issue analyzed in the story and explored in the exercise. While the exercise pushes the student to relate practically to the general media issue, the analysis criteria propose an abstract and theoretical point of view. That's why the debate proposed must be linked to the media literacy skills suggested by the analysis criteria so that, through them, students can go deeper into media structures that arise from the story.

A framework was developed to address media literacy around constructs that can be extrapolated to media systems and cultural ideologies around the world. These questions address the 5 A's of media literacy–**access** to media, **awareness** of media's power, **assessment** of how media cover international events and issues, **appreciation** for media's role in creating civil societies, and **action** to encourage better communication across cultural, social and political divides, (Mihailidis, 2009), these constructs collectively support existing media literacy skills and learning outcomes (Mihailidis, 2009). The first three A´s–access, awareness, and assessment–advance the student's critical thinking about the media while the last two–appreciation and action–encourage the understanding of their relevance for the healthy development of civil society and the active engagement of students in civil society. Wrote Mihailidis (2009):

> "Media literacy education should cultivate students who can effectively read the media. This entails greater critical analysis skills (comprehension, evaluation, assessment), critical thinking skills (awareness, reflection, engagement) and an appreciation of the necessity of media for civil society. In this way, media literacy education can go beyond basic media and communication courses in that although it is grounded in inquiry-based pedagogy, it provides "a new way to teach and more importantly, a new way to learn" (Thoman & Jolls, 2004)." (p.65)

Questions about **access** approach issues such as the digital divide, the participation gap, barriers to information and access to information as a fundamental human right. For as the rise of new media technologies has allowed for mass information flow globally, access to information moves to the forefront of individual rights (Mihailidis, 2009). Students should be asked to describe these topics and also to debate their influence on individuals and society at large.

Example 5:
Case Study: Bridging Religious Divides—The 2006 War between Israel & Lebanon
By: Muneira Hoballah and Su Jing

ACCESS
When a country is in conflict, should the citizens have access to the opposition's media?
What happens when that media is in a different language?
Should citizens be able to access information and viewpoints from new forms of media?
In Lebanon, it is illegal to communicate with an Israeli due to the nature of the countries' political relations. How does that law affect the public's ability to access viewpoints and better understand the other's viewpoints?
How does making it illegal to communicate with an 'opponent' affect the sentiments of the public and current as well as future relations with the opposing parties and citizens?
Is access to a third party's media enough?
How do religious beliefs, foreign relations, and/or the type of political system you live under affect your access to media from 'the opposition'?

Awareness questions disclose the power of particular discourses, and the conditions in which media messages become powerful. Students should ask themselves "what is the meaning of the information provided in each case in larger social and civic contexts; what are the main issues in the information presented; what are the underlying assertions; how are the stories being told, and by whom" (Salzburg Academy on Media and Global Change, 2008). Being aware of media's power necessarily entails understanding the complexity and cultural context of media messages. This is seen as a fundamental attribute for considering the values associated with information from its point of origin.

Example 6:

Case Study: Stereotyping and National Identity in Entertainment Media

By: Nancy Azar and Simon Essex

AWARENESS

Television shows and movies are an important window into another culture or nationality. Do you think that there are better ways, better media for making people aware of other nationalities, religions or cultures? To what extent do entertainment shows challenge or reinforce national stereotypes?

Assessment questions encourage students to analyze the representation of conflicting issues. They should discuss, "who is the intended audience in a particular report or program; what are symbols; what are hidden messages; from what angle is the story being told; from what other angles has it been told; what's the emotional appeal; who is speaking, delivering the message; and specially what is omitted from the message" (Salzburg Academy on Media and Global Change, 2008).

Example 7:

Case Study: Bridging Religious Divides~The 2006 War between Israel & Lebanon

By: Muneira Hoballah and Su Jing

ASSESSMENT

Assess how mainstream media outlets from different countries cover conflict, especially any religious aspects of a conflict. You might want to look at the coverage of conflicts in the Middle East, in South Asia, in Sudan, or elsewhere.

Note how blogs and other new forms of media, such as YouTube or Tudou, have gained importance as news outlets. Why has this happened? Why are people, recently, turning towards these outlets?

Consider the true story above. Assess how the media, in this case email, helped to foster understanding between the two girls during the war of 2006. Would their growing understanding of each other have been different if they had used another form of media?

In the **appreciation** analysis criteria students are expected to debate how access to diverse, independent, and accurate information is beneficial for the development of a healthy civil society. They should also be able to discuss which types of media texts, if any, are detrimental to civil relations, and if so, should be limited. Are violent images necessary or should their publication be limited or even stopped?

Example 8:

Case study: The Pulling Down of Saddam Hussein's Statue in Baghdad, April 9, 2003

By: Tanya Kassab and Lorena Figueroa

APPRECIATION

Exposure to news and information and different points of view allows the public to get acquainted with notions of civil society and of responsibility. Do you think pictures and video clips "capture" information that a story or article does not necessarily present?

Do you think that pictures and video clips might not present certain information needed for a good understanding or a story or event?

Do you feel people appreciate the power of images and are able to grasp what images are trying to express?

Can images help create civil societies and if so, in what way?

Action encourages the promotion and protection of the opportunities to develop positive kinds of media discourses. Mihailidis (2009) writes, "Never before have there been so many avenues for active participation in global dialogue as there are now. Internet and new media technologies have enabled new means for media production and activism. Global media literacy must teach how such newfound avenues for expression can empower people to take action." In this sense, action entails the positive contributions—large or small—of individuals to civic dialog that helps ensure a diversity of voices, a tolerance of the other, and a vibrant discourse between local and national and global communities.

Example 9:

Case study: Chávez and the RCTV Shut-down

By: Celia Scheff and Natalia Bocassi

ACTION

How can images of a news event help an audience to care or even to take action about that issue? Should journalists themselves take action on issues they care about? Do the choices that photographers (and reporters) make of what to cover constitute a kind of action? Should there be a difference between how professional photographers and civilian photographers (say with a camera phone) act—take photos—when they are in the midst of an event that is newsworthy. Can photographers—either professional or civilian photographers—go too far to document an event? Citizens often participate in demonstrations but then turn to take pictures of the event. Is it necessary to monitor the action of these citizens?

Example 10:

Case Study: Xenophobic Violence and Press Coverage in South Africa in 2008

By: Hebresia Present and Rachel Leven

ACTION

What kind of action can you take to ensure that especially children are least exposed to these images and racial stereotyping?

What media actions do you think should be monitored? What types of response from readers and viewers would be appropriate?

Would you consider actions like writing to editors, starting blogging sites or even directly challenging reporters of these articles?

Conclusion: Creating Shared Dialog

Scholarship on media literacy has been prolific in explanation and debate about its scope, its relevance, the dilemmas brought forward by the digital age and the possibilities of global communication. The exploration of the most appropriate teaching techniques to successfully achieve these outcomes in a global context is harder to find. This void leaves teachers looking to teach about global media with a long list of contents to cover and skills to provide to students, but with little help on how to achieve those pedagogical goals.

This chapter has sought to fill this gap through the description and analysis of case studies as a classroom technique. The elaboration of case studies, as developed by participants in the Salzburg Academy on Media and Global Change, has proven to be an efficient media literacy learning opportunity for their final users, but also for their creators. In the construction of stories that address relevant media issues, exercises and analysis criteria, students are driven to research, analyze, debate, experiment and to produce new content for other students. The diversity of teaching tools they must produce is directly linked with media literacy skills and learning outcomes previously described in this chapter.

Research for the story demands that students critically view one particular media text, and relate it to a broader issue. The discussion of findings and the writing of the story challenge their preconceptions and forces them to address the diversity of points of view that surrounds every media problem. The hands-on experience of an exercise challenges students with the real difficulties and innuendo involved in abstract discussions. Finally, the construction of analysis criteria linked to the 5 A's of media literacy promotes a kind of debate that avoids the risk of promoting cynicism by encouraging critical thinking about the media, and also the intent to value and protect its role in civil society.

Through the elaboration of learning modules students are simultaneously engaged critically and emotionally to abstract and complex media issues. They are required to go past their own prejudices into broader media aspects through debate, research, application and, finally, by repurposing conclusions into new questions. These activities can be simultaneously understood as a structured pedagogical tool, with very clear parts and processes, oriented by very clear learning outcomes, as a well as a very creative and flexible process that allows students to conduct much of their own learning by investigating their own topics, producing their own contents, and reaching their own conclusions. This, together with continuous renewal of these modules by new generations of students, the use of them in diverse classroom contexts, and the sharing of them through digital media, can provide a unique and "global" approach to media literacy that is student driven, collaborative, and connects critical analysis to understanding media's role community voice, democratic progress, and civic empowerment in the 21st century.

REFERENCES

Aufderheide, P. & Firestone, C. (1993). *Media Literacy: A Report of the National Leadership Conference on Media Literacy.* Cambridge, UK: Polity Press.

Azar, N., & Essex, S. (2008). *Case Study: Stereotyping and National Identity in Entertainment Media.* Retrieved 2010 December-21 from Salzburg Academy on Media and Global Change: http://www.salzburg.umd.edu/salzburg/new/lessons/602

Buckingham, D. (2003). *Media Education: Literacy, Learning and Contemporary Culture.* Cambridge, UK: Polity Press.

Castillo, A., & Almeida, W. (2008). *Case Study: Eco-Journalism & Cleaning Up Sumidero's Canyon.* Retrieved 2010 December-26 from Salzburg Academy on Media and Global Change: http://www.salzburg.umd.edu/salzburg/new/lessons/506

Duran, R. L., Yousman, B., Walsh, K. M., & Longshore, M. A. (2008). Holistic media education: An assessment of the effectiveness of a college course in media literacy. *Communication Quaterly,* 49-68.

Hobbs, R., & Frost, R. (2003). Measuring the acquisition of media-literacy skills. *Reading Research Quaterly,* 330-355.

Hobollah, M., & Jing, S. (2008). *Bridging Religious Divides - The 2006 War between Israel & Lebanon.* Retrieved 2010 21-December from Salzburg Academy on Media and Global Change: http://www.salzburg.umd.edu/salzburg/new/lessons/321

Kassab, T., & Figueroa, L. (2008). *Case Study: The pulling down of Saddam Hussein's statue in Baghdad, April 9, 2003.* Retrieved 2010 26-December from Salzburg Academy on Media and Global Change: http://www.salzburg.umd.edu/salzburg/new/lessons/107

Leibowitz, M. (2009). *Case Study: The power of citizen journalism demonstrates the need for standards.* Retrieved 2010 26-December from Salzburg Academy on Media and Global Change: http://www.salzburg.umd.edu/salzburg/new/lessons/1001

Mihailidis, P. (2008a). Are we speaking the same language? Assessing the state of media literacy in U.S. higher education. *Studies in Meidia & Information Literacy Education,* 1-14.

Mihailidis, P. (2008b). Beyond cynicism: How media literacy can make students more engaged citizens. *Report for the Salzburg Academy on Media and Global Change* . Salzburg, Austria.

Mihailidis, P. (2009). The first step is the hardest: Finding connections in media literacy education. *Journal of Media Literacy Education,* 53-67.

Oxstrand, B. (2009 йил 13-15-August). *Media Literacy Education: A Discussion about Media Education in the Western Countries, Europe and Sweden.* Karlstadt University, Sweden.

Present, H., & Leven, R. (2008). *Case Study: Xenophobic violence and press coverage in South Africa in 2008.* Retrieved 2010 26-December from Salzburg Academy on Media and Global Change: http://www.salzburg.umd.edu/salzburg/new/lessons/201

Salzburg Academy on Media and Global Change. (2008). *Global Media Literacy: A New Curricula.* Retrieved 2010 йил 16-December from http://www.salzburg.umd.edu/salzburg/new/media-literacy-curricula

Scheff, C., & Bocassi, N. (2008). *Case study: Chávez and the RCTV shut-down.* Retrieved 2010 26-December from Salzburg Academy on Media and Global Change: http://www.salzburg.umd.edu/salzburg/new/lessons/202

Tisdell, E. (2008). Critical media literacy and transformative learning. Drawing on pop culture and entertainment media in teaching for media diversity in adult higher education. *Journal of Transformative Education* , 48-67.

Uriarte, M. (2009). *Case Study: The Cloning Fraud: How Can Media and Science Live Together?* Retrieved 2010, 26-December from Salzburg Academy on Media and Global Change: http://www.salzburg.umd.edu/salzburg/new/lessons/6003

Chapter 6 -
The Role of Multimedia Storytelling in Teaching Global Journalism: A News Literacy Approach

MOSES SHUMOW

Florida International University, Miami, USA

SANJEEV CHATTERJEE

University of Miami, USA

Introduction

From across the rapidly transforming field of journalism, there is a growing emphasis being placed on university journalism programs to arm their graduates with an increasingly large and diverse set of multimedia production and storytelling skills (Downie & Schudson, 2009; Huang, 2009; Presley, 2010). New media technologies have become cheaper, easier to use, and more accessible, while at the same time the quality of these products has continued to rise. However, given the relative newness of the push towards media convergence in the newsroom (Dupagne & Garrison, 2006; Quinn, 2005), and shifting definitions of what it means to be a journalist in an era of crowd-sourcing, social networks, blogospheres, and Twitterverses, there remain many pedagogical questions surrounding how to most effectively teach these new forms of storytelling. What are the most effective teaching methods for multimedia reporting? What sort of classroom environment is most conducive to obtaining these skills while at the same time encouraging the sort of critical engagement with the subject matter that will lead to effective, compelling journalism? How will students respond to these new forms of reporting that reach across media platforms and break traditional silo models of journalism?

This chapter begins to address these questions by examining two multimedia projects conducted during the 2009-2010 school year by students in the School of Communication at the University of Miami, USA. At the heart of this approach, and the element that we believe makes this contribution unique, is the application of a media literacy framework for teaching multimedia storytelling. This approach emphasizes the attainment of multimedia abilities not only on a professional level, but as a way of teaching students how to navigate an increasingly complex digital, globalized media landscape as both a producer (in the role of journalist) and critical consumer (in the role of citizen) of media. It also envisions the transformation of the traditional learning model of knowledge *transfer* from teacher to pupil into a process of knowledge *gathering* through practice, engagement and collaborative discovery as the most effective model for approaching these new forms of storytelling.

Drawing on conversations with instructors and feedback from students, we argue that teaching multimedia storytelling must push students to go beyond traditional conceptualizations of what it means to be a journalist: the detached observer delivering a balanced account of reality. While the need for these traditional models has not disappeared, multimedia allows us to rethink the "objective omniscience" model of the 20th century and envision a form of journalism that forces media producers to transition from a hands-off, "high-priest" model of the journalist to a model in which the reporter actively and collaboratively engages with the subject and audience. Indeed, the interactivity afforded by the Internet has profoundly transformed notions of journalism. Teaching this new model requires a three-pronged approach, a strategy that at a minimum should entail the following:

1. Students must embrace engagement in the classroom ("learning by doing") and with the audience. The ways in which journalism students can now collaborate and interact with their audience creates new opportunities for transforming the classroom into a dynamic environment. This step is facilitated by the accessibility of new digital technologies that allow for less time spent on teaching technological competencies and a greater focus on developing a critical understanding of storytelling, subject matter and audience.

2. A renewed emphasis on audience interaction must also incorporate a "classroom without walls" strategy in which students engage in the world around them. This means finding new ways for students to give voice to their subjects through collaboration, teamwork, and topic negotiation, as well as taking full advantage of the interactivity made possible by multimedia platforms that include social media and network building.

3. Students must be pushed to develop a deeper understanding and appreciation of the tools that make up a multimedia project in order to maximize the effect of each element in telling a story that informs, compels, and entertains. Multimedia incorporates multiple storytelling components – audio, video, photography, text and infographics – and it is imperative that students grasp the nature of each and can identify appropriate applications.

At the same time, we argue that these student projects, because they are taking place within higher education, should take on an international, multicultural orientation. In an interconnected world, in which graduating university students are more likely than ever to come into daily contact with different cultures, perspectives, and ideologies, as well as transnational problems that ignore borders (e.g., terrorism, climate change, outsourcing, global economic recession, trafficking in drugs, humans, and arms), institutes of higher education are increasingly being pushed to redefine their role to emphasize producing students that are global citizens (Gibson, Rimmington & Landwehr-Brown, 2008; Howland, 2006). By focusing on projects with a global perspective, such as those that will be outlined and explored below, students are pushed to seek out the connections between media production, globalization, multiculturalism, and the role of media producers in both shaping and reporting on these processes. Such multimedia storytelling allows students to engage with the nuances of a globalized world in new and dynamic ways.

Two Global Multimedia Storytelling Projects

The first reporting project that will be examined in this chapter highlights a communication honors course that explored the lives of young people in Miami and Havana, fifty years after the Cuban revolution. The "Havana-Miami"[1] project revolved around a series of two-minute video vignettes offering glimpses into the day-to-day lives of twelve young people – six Cubans living in Havana and six Cuban-Americans living in Miami. The videos, along with biographies and interactive timelines of the lives of the participants, were all hosted on a multimedia website designed, developed, and promoted by Arte (Association Relative à la Télévision Européenne), the Franco-German television network. Both authors of this chapter, along with a third instructor, Mark Mochabee, were involved in the teaching and execution of this project. Students from the class were also asked to fill out a brief online questionnaire (outlined below in greater detail), in order to gain insight into their experience working on the class project.

The second project entailed a semester-long investigation into the United Nation's Millienium Development Goals (MDGs)[2] by a group of graduate students enrolled in the multimedia journalism program of the School of Communication, working under the tutelage of Rich Beckman and Tom Kennedy,[3] both recognized leaders in the field of multimedia. During the Spring 2010 semester, fourteen students working on this project, in teams of two, travelled the globe gathering multimedia stories that were connected with the MDGs, from the status of maternal health in Freetown, Sierra Leone, to environmental stability in Durban, South Africa. Similar to the "Havana-Miami" project, the results of these students' efforts were built into an interactive, multimedia website called MyStory, MyGoal,[4] allowing users to navigate between the seven stories gathered, watch all of them in a feature length documentary, as well as learn more about the MDGs and where the world stands in addressing these critical global issues that confront humankind at the beginning of the 21st century.

This chapter is built on the findings resulting from a close examination of these two projects, drawing on conversations with both students and instructors (in the form of both the online questionnaire mentioned above, and in-depth interviews with Beckman and Kennedy). We use the pedagogy of digital media literacy as a theoretical lens for analyzing the work and experiences of the students as they tackled two ambitious multimedia storytelling projects. [Literature on enabling students with "new media literacies" (Jenkins, 2006; O'Brien & Scharber, 2008) through the teaching of production skills along side forms of critical-analytical thinking argues that media production skills and the ability to take part in non-traditional, multimedia forms of articulation must be seen as key to students engaging in new platforms for digital storytelling]. The tenets of this new form of literacy tell us that critical thinking and an ability to navigate increasingly complex media environments are an essential element in fostering a democratic citizenry (Buckingham, 2003; Hobbs, 1998).

While analysis and reflection represent one part of this equation, increasingly accessible digital tools are providing new opportunities for students to produce rich multimedia content for distribution and discussion beyond classroom walls. These are skills that were limited to a small group of media elite just a few years ago. Today, easy access to video cameras and other recording devices, coupled with simple editing programs, allows for innovative approaches to student engagement through media production. In addition, there is a growing recognition that today's generation has new levels of digital proficiency that were previously unimaginable; early and constant exposure to digital technologies has created a new cohort of "digital natives" (Palfrey & Glasser, 2008) who have "[m]ajor aspects of their lives – social interactions, friendships, civic activities... mediated by digital technologies" (2). Thus, as digital technologies become ever more ubiquitous, instructors now have the opportunity to spend less time on teaching skills and more time emphasizing critical, engaged learning that forces students to think analytically, tackle challenging topics, and develop the abilities needed for innovative problem solving.

Each of the projects outlined here immersed students in the skills needed for multimedia production–video, photography, blogging, web-building, social-networking–teaching them to draw on the strengths of each medium, a process that is fundamental to creating effective interactive media. At the same time, each of the projects enabled students to find ways of understanding the world and global issues through the personal stories of individuals. In this regard, these efforts fall within previous definitions of *digital fluency*, with students "not only knowing how to use technological tools, but also knowing how to construct things of significance with those tools" (Resnick, 2002, 33).

The students were also encouraged to see their work within a larger global context. By exploring opportunities for audience participation that moved beyond traditional sender-receiver models of mediated communication, these multimedia projects encouraged a continuous dialogue with both the subject matter and intended viewers through a deep engagement with the material and their collaborators, blogging about their experiences, and utilizing social networks for interacting with the projects. In some cases, it was even possible to have students analyze aspects of the online audience and its preferences through the use of tools like Google Analytics. The end result was both the creation of innovative and unique forms of storytelling and an in-depth and cross-cultural engagement on the part of the students with issues of global scope and impact.

The Pedagogy of Multimedia Production

Among the multiple and varied definitions and interpretations for what constitutes media literacy,[5] the basic fact remains that literacy, in *any* form, advances a person's ability to effectively and creatively use and communicate information (Jones-Kavalier & Flannigan, 2006, 9). By approaching the peda-

gogy of multimedia production from a media literacy perspective, we are drawing on the definition outlined by O'Brien and Scharber (2008), who define
digital literacies as "socially situated practices supported by skills, strategies, and
stances that enable the representation and understanding of ideas using a
range of modalities enabled by digital tools" (67). Making the connection between media production and digital literacy allows us to approach this subject
with a broader, enhanced perspective about the meaning of a journalism degree in an era of economic, political, and cultural globalization. This will be
essential for students graduating into a media environment that will both challenge them to be innovative and enterprising as they pursue their careers, and
at the same time offer more opportunities for unique forms of information
gathering, data visualization, reporting, storytelling, and audience analysis,
than at any other point in the history of the profession.

In his book on the complex landscape of 21st century media cultures, *Convergence Culture*, Henry Jenkins (2006) writes:

> The American media environment is now being shaped by two seemingly contradic-
> tory trends: on the one hand, new media technologies have lowered production and
> distribution costs, expanded the range of available delivery channels, and enabled
> consumers to archive, annotate, appropriate, and re-circulate media content in power-
> ful new ways. At the same time, there has been an alarming concentration of the
> ownership of mainstream commercial media conglomerates dominating all sectors of
> the entertainment industry. (8)

Today's journalism students are facing this new environment with the
added pressures of having to navigate these new forms of convergence for reporting while entering a profession under of intense economic and professional uncertainty. If approached in a holistic way that combines skills with
critical thinking and audience engagement, multimedia storytelling can give
students an added set of tools to help them launch their careers.

However, media educators must be sure that they emphasize the need for
the tools of production to move beyond specialization. This entails, as Burns
and Durran (2006) have written in *Digital Generations*, that teachers draw the
connection "between the new possibilities for production and the analytical and
interpretive work of media education" (275). In this light, and further embracing the point made by Jenkins, media production cannot and should not be
separated from a critical understanding of the media. By pushing students to
give voice to their subjects, work collaboratively, approach topics from a global
perspective, and enable outcomes that have tangible impacts beyond the class-

room, students can attain the analytical and technical skills we see as essential to effective communication through multimedia production. At the same time, instructors should approach this kind of teaching as a shared exploration with the students, following the advice of Resnick (2002): "Teachers cannot simply pour information into the heads of learners; rather, learning is an active process in which people construct new understandings of the world around them through active exploration, experimentation, discussion, and reflection" (33).

The Miami-Havana Project

Image 6.1: The Miami-Havana project. Courtesy Authors.

The "Miami-Havana" project was the outcome of a collaboration between the Knight Center for International Media, housed at the University of Miami's School of Communication, and Tamouz Productions, a New York-based documentary production company. During the summer of 2009, Tamouz founder and CEO, Ilan Ziv, approached Knight Center Director (and this chapter's co-author) Sanjeev Chatterjee to explore the possibilities for engaging students in a semester-long, multimedia-reporting project that would tell the stories of the day-to-day lives of young people in Miami and Havana. Ziv had extensive experience in this sort of storytelling, having produced similar projects in the past in Palestine, Israel, and Europe. Teams of multimedia producers in Miami and Havana would find 12 characters (six in each city) and follow their lives over several months, chronicling each chapter in two-minute web videos that would be rolled out on a week-to-week basis, allowing audiences to follow the characters. Chatterjee saw the goals of the project as being very much in line with the Knight Center's goal of producing "compelling visual media crossing borders to solve the world's most difficult problems,"[6] and set up a communication honors course to tackle the project during the fall of 2010.

Students in the class, under the guidance of Chatterjee and two graduate students, were responsible for the entire Miami portion of the project: they had to find the characters, conduct pre-interviews, plot out the six "webisodes" that would be presented as snapshots of their lives, set up and execute the filming, and edit the six two-minute videos for each of the Miami-based characters (a separate Cuban team in Havana was responsible for the Havana stories). They also made regular blog entries on a class website, in which they reflected on the process as it was taking place, maintained a Facebook page for the class, and conducted background research on Cubans in Miami. It is important to point out that the students in this class had very limited levels of experience in multimedia production and the kind of character driven storytelling that they were being challenged to produce; while some had experience in non-linear video editing and field production, most had never touched a video camera. These were not only journalism majors; they were also public relations, advertising, media management, and electronic media majors. However, driven by the belief that the best approach to this project was to "learn by doing," the instructors first surveyed the class to get an idea of student skills-sets and interests, then spent a week conducting intensive production workshops on gathering quality video and audio, and the basics of video editing. The students were then set loose in the city to start finding young Cuban-Americans whose lives they would chronicle over the next four months.

Engagement and collaboration were fundamental to the structuring of this course. Working under tight deadlines (each group of 3-4 students had to find their character and produce six two-minute segments during the course of the semester), students were obliged to work closely together, discovering each other's strengths and weaknesses and immediately begin seeking out the character whose stories they would tell. Finding the right people to feature in the project was a fundamental challenge. While some pre-production work had been done prior to the start of the semester, and some characters had been approached initially, it was up to the students to pre-interview the character, drawing out details of their lives and building rapport. They then had to develop an outline for how each of the six two-minute episodes would play out on the website; each episode had to have its own story arc, and also needed a connecting thread and recurring themes that would hold all of the stories together. The resulting characters and their stories were diverse and wide-ranging, from a Columbia Law-educated third generation cigar maker launching a new company with her father and brother and finding her way in a male-dominated industry, to a struggling musician living with his grandmother; a high-school theater student with dreams of Broadway and a family fighting foreclosure on their home; and a former Olympic gold-medalist wrestler making a name for himself in the brutal world of professional mixed martial arts.

As instructors, we took a very hands-off approach in teaching the class, instead following the "knowledge acquisition" and "learning through doing" pedagogical models outlined above. There were a limited number of lectures, including a discussion of *digital fluency*, using Mitchel Resnick's (2002) article on "Rethinking learning in the digital age" as a starting point, in-class feedback and group dialogue on what it means to live in the city of Miami, perceptions and knowledge of the Cuban-American community, and what the students were learning about the city as the production progressed. However, all three instructors were primarily there to serve as guides, sounding boards, and reference points for the students as they discovered and gave form to their characters' stories.

Interested in drawing on the experiences of the students in order to fully integrate their perspective into this chapter, we asked the students to fill out a brief, online survey, made up of the following open-ended questions:

1. Did your understanding of media production change over the course of this class? If so, how?
2. What were the greatest challenges you faced in learning the skills and tools needed for producing a multimedia storytelling project?
3. Did gaining those media production skills change your perception of how media are created and the way they function?
4. Did your understanding and perception of the city of Miami change during the course of the project?
5. Do you believe multimedia projects like Havana-Miami should be incorporated into wider classroom environments?

We developed these questions with the goal of exploring the basic tenets that we felt were needed to make a class like this work: the importance of engagement and collaboration; pushing students to learn through experience; and encouraging them to seek out a deeper understanding of media production within the context of a globalized, multicultural setting.[7]

In their responses to Question 1, examining changes in their understanding of media production, the students responding to the survey appeared to confirm the argument that a course like this one, which pushes active collaboration, can encourage students to develop a broader, deeper understanding of producing effective storytelling. Wrote one student, "I learned about all that goes into the process of making a documentary, including the hardships of scheduling, getting the scenes you want, post production, etc." This outcome was echoed by another student who wrote, "I had never been exposed to all the aspects of media production. I knew it took a lot of work, but seeing all the hours that went into such a short segment really increased my appreciation for this field."

The students responding to the questionnaire clearly found both coming up to speed with the technology and producing content for the web along with creating character-driven storytelling to be the most challenging aspects of the class. "The greatest challenge was trying to find an interesting story that would hold the audience's attention for two minutes," wrote one student. "Because we could not shape or create what we wanted, it was much more difficult to initially see what the most compelling arc was." The way the class was structured, with multiple levels of feedback from in-class screenings with classmates, the instructors, as well as from Ilan Ziv at Tamouz Media, and even executive producers at Arte, the project's European partner, meant that students were continuously reworking their edited pieces in an effort to address multiple comments and draw out the most compelling elements. The process highlights the centrality of collaboration during the production of these projects. Despite their frustrations, the process of developing characters, finding more room for the central character in each piece to tell their own story, and working with multiple levels of feedback and criticism from fellow students, instructors, and producers, pushed the students to engage with the material in a way that would be difficult to recreate in a traditionally structured news reporting class.

Similarly, students were sometimes frustrated by the unfamiliar technology, but ultimately rose to the challenge of learning these tools on the fly. While the students were not always thrilled with this "sink or swim" approach, the following comment illustrates the way in which collaborative work can push team members to find the areas where they are strongest and can make the most solid contributions to the effort:

The biggest challenge I faced was that I had no background experience. It was probably not a class I would have elected to take because of this. Ultimately, I was given the
task of creating the webpage. While I had to learn entirely new platforms to create
the webpage, this task was more suited to my individual skills.

This feeling was echoed by a second student who wrote, "My teammate and I
had to essentially teach ourselves web language and how to master the art of
the web." The pressure of meeting a deadline along with the rest of the class
meant that these students had to step out of their comfort zones and enter
into unfamiliar territory, learning as the project progressed.

Building on the idea that these projects, because they take place in a university setting, should give students a more global perspective of the topic they
are covering - in this case Miami and the long history and interwoven cultural
and political ties that the city shares with Havana, Cuba - we wanted to know
in what way this class had changed the students' views of the city in which they
studied and lived. All of the answers to Question 4 ("Did your understanding
and perception of the city of Miami change during the course of the project?")
revealed a marked change in perceptions of the city. This result is perhaps best
summarized in the following commentary, which is an insight into the insular
lives that students can live when in university, but also the way in which exposure to new experiences and lifestyles can disrupt that isolation: sometimes, as
a student, I think of Miami as a campus and a nightlife scene. I forget that
people are born here, people migrate here, people grow up here. I forget that
everyone has a story, and that outside of the glitz and glamour of South Beach
and the college vibe of other areas, there are many, particularly minorities,
who are struggling. One student described, "feeling a connection to them (the
characters) by constantly seeing their stories and working with footage collected with them," while another wrote that, "Miami is even more diverse than
I thought before. Getting to see personal stories from the area showed me all
the different types of people that live in this area outside of the students."

If universities in the 21st century are to stay relevant as spaces for developing global awareness in their students and producing world citizens, then this
comment from a "Miami-Havana" student would seem to truly encapsulate the
direction in which these changes should migrate. Gibson, Rimmington, and
Landwehr-Brown (2008) highlight this new direction in higher education in
their article, Developing Global Awareness and Responsible Citizenship with
Global Learning: "Leaders in a globalized world need skills that allow them to
collaborate, communicate, negotiate, think critically, and gain multiple per-

spectives through dialogic co-construction of meaning with individuals from different cultures" (12). The above reflections from "Miami-Havana" students would seem to indicate that, at least to a certain degree, some of these skills had been introduced to them through their work in this class.

My Story, My Goal

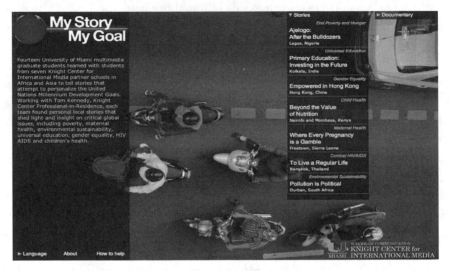

Image 6.2: My Story, My Goal. Courtesy of Authors

"My Story, My Goal" grew out of a proposal from the Knight Center for International Media resident professional Tom Kennedy. Kennedy, former managing editor for multimedia at Washingtonpost/Newsweek Interactive, approached the Center with the idea of structuring a graduate course around reporting on the progress of the Millennium Development Goals (MDGs) that were developed by the United Nations in 2000. Kennedy teamed up with Knight Chair in Visual Journalism Rich Beckman. Following an introductory semester in which the fourteen students who took part in the project spent time honing their production and storytelling skills on a variety of smaller projects, the students were paired into teams of two and assigned one of seven MDGs.[8] Through previous work by the Knight Center in building worldwide partnerships with journalism and media programs around the world, the students began to conduct intensive background research and develop stories in collaboration with students at partner institutions throughout Asia and Africa.

They were faced with the challenge of finding ways to give a face and voice to the progress of the MDGs around the world.

The final result of this work is the interactive multimedia website, "My Story, My Goal." Over the course of a semester, students in the class traveled to Lagos, Nigeria; Kolkata, India; Hong Kong; Freetown, Sierra Leone; Bangkok, Thailand; Durban, South Africa; and Nairobi and Mombasa, Kenya, gathering material for their reports, and covering stories ranging from child health in Mombasa and Nairobi to battling the HIV/AIDS epidemic in Bangkok. Through the stories of individuals, each video on the site puts a human face on the MDGs, and, similar to the "Miami-Havana" webisodes, the stories avoid any third-person narration, allowing the voices and stories of those featured to drive the story forward. Visitors to the site can view each story on its own or all of the pieces as a single, long-format documentary; photo slide shows add further depth to each character and story and infographics on the site offer more in-depth reporting on the state of the MDGs worldwide.

Interested in gaining further insight into the teaching multimedia projects like "My Story, My Goal," especially from two leaders in the pedagogy of multimedia production, the chapter's co-authors sat down with Professors Kennedy and Beckman to ask them about their experiences in teaching this class. Several key themes emerged from the interview, each very much in line with the outcomes of the "Miami-Havana" project. During the conversation, it emerged that team building was an essential element in setting up the class. As Beckman related, by the time the project started, even though the students "all had had enough experience" from projects the previous semester:

> "... there is no doubt they were at different levels. Part of that was addressed by meeting with the more experienced students and deciding how to assemble the teams....We teamed them up not only based on skills but also their personalities and ideologies, as well as where we were going to send them. There might be students in a class that are weaker technically, but very good with thinking about things. Then there are always students who can never work with each other, even though as individuals they may be great....You've got to deal with these things...this is no different that a newsroom - this is a multimedia newsroom."

Throughout the class, the interaction and back-and-forth involvement between students and instructors was intense. Thoroughly researching and understanding the MDGs and challenges being faced by people living in the cities under study was an essential part of the class. Because these were tight shooting schedules in unfamiliar and challenging locations, the teams also had to carefully think through the right places to go and the best stories to cover, given the goals of the project and the multimedia formats in which they would be working. Then on the back end, in editing the individual stories as well as the overall documentary, Kennedy related how the students were very "unused to" having a faculty member work hands on as much and as extensively as he did. However, despite the challenges, Kennedy found the experience satisfying because he "got to take them all the way through the process." Adopting a similar model used in "Miami-Havana" of learning through doing, Kennedy tried to "let the students lead...we gave them a basic framework operationally for what we wanted to achieve" but in the end, there was a "certain burden of learning on the students and asked them to take a certain kind of responsibility for their own growth."

Beckman related how intensively he worked with the students prior to production in order to set up the stories in advance. Project design was essential to a successful production, and this meant communicating closely with partners at the universities in the countries where the stories would be filmed. The American students relied heavily on their partners in the cities where they would be filming to help find the characters, set up the stories, and arrange for filming, as well as help deal with issues relating to language and translation. The challenges in this process were multiple, from dealing with disagreements and deciding on a style of storytelling to figuring out the actual content of the stories, and in the end, finding a way to arrive at a mutual decision.

A further challenge lay in the fact that students were using video, photo galleries, infographics and, to some extent, different languages. However, this also allowed them to think about the most appropriate medium for the stories they were telling. Said Kennedy: "They had to look at multimedia journalism as if they had everything in their tool palate...here is the story, figure out the best tools to tell it...we talked a lot about how different parts of the story could be best told with different tools; how to bring it together in a cohesive design that was usable." For Kennedy, who served as story editor during post-production, it was essential to get a firm understanding from the students, once they were back, as to what they thought they had in terms of material. This meant interacting with students on a daily basis, working with the stories and helping students hone the stories so that the voices that best told the story would come through. It was a time-consuming process, but ultimately rewarding. "I just wish we could teach all our classes that way," Kennedy said.

Finally, in the eyes of both Beckman and Kennedy, a course like "My Story, My Goal" appears to be a valuable avenue for encouraging students to build a critical understanding of globalization and the role of media in developing diverse multicultural perspectives. As Beckman put it, students in a course like this one "learn about their own media landscape in relationship to the media landscapes in the rest of the world – both from a storytelling/audience perspective and the technology (digital divide) perspective." Kennedy added that a project like "My Story, My Goal" allowed students to "discover where their passions and skill sets lay. At the same time, they find out the power of collaboration to leverage each others' strengths... the crucial thing is the collaboration as opposed to the silo approach...the challenge is that most media companies still don't understand that."

Conclusion: The News Literate Multimedia Storyteller

As a way of concluding, it is important to point out that both "Miami-Havana" and "My Story, My Goal" had the advantage of being taught by seasoned instructors with heightened access, through the Knight Center for International Media, to technological resources that allowed the students to put together strong multimedia projects that would hold up against most professionally-produced projects. Additional support from the Knight Center allowed for extensive traveling during the production phases of "My Story, My Goal." While the costs of multimedia production equipment have declined significantly in recent years, giving more students on a wider array of campuses access to these tolls, not all universities are as fortunate to have the kinds of resources upon which these two projects were able to draw.

However, the goal of this chapter has been to draw on these projects in order to propose a more universal pedagogical framework for teaching multimedia production and storytelling, based on the tenets of media literacy and global journalism, with the hope that media instructors worldwide might duplicate such approaches in their own classrooms. As teachers and researchers, we hold no illusions as to the challenges inherent in any attempt at this type of classroom restructuring. The level of students' enthusiasm for the collaborative process, for tackling technical hurdles through creative problem solving, and for engaging more thoroughly in the subject matter will always place limits on serious, ambitious multimedia projects. Teachers will also have to confront their own preconceived notions about their role in the classroom, finding a willingness to relinquish some control over the structure of the course while at the same time remaining constantly vigilant as to the progress of their students and stepping in to guide them in the right direction when needed. However, the pay-off that is gained through the self-learning and a new, critical understanding of what it means to produce and share meaningful stories through multimedia production is at the forefront of what educators strive to achieve in the 21st century.

At a time of profound uncertainty about the state of journalism and the directions in which the profession is headed (Bird, 2009; Downie & Schudson, 2009; Quinn, 2005) , constantly shifting definitions for what it means to be a journalist (Deuze, 2004, 2005), and the larger implications for the future of media industries as well as the functioning of modern representative liberal democracies (Barnett, 2002; Dahlgren, 2003), multimedia storytelling represents one of the clearest indicators of the future of the profession. In an age of media convergence, journalism students are going to need multimedia reporting skills in order to survive. The key for instructors will be providing those skills in a way that is collaborative, critical, hyper-contextual, and moves away from silo models of reporting and embraces the potentials for new forms of reporting and storytelling that multimedia production represents.

Concurrently, as the goals of achieving a higher education are also being re-imagined for relevancy in a global context (Banks, 2007; Gibson, Rimmington & Landwehr-Brown, 2008; Howland, 2006), it is possible that intensive, semester-long multimedia projects could serve as a catalyst for harnessing and combining the energies of both the creative and research communities that help make American universities so dynamic and among the best in the world. The background research that went into the pre-production of both "Miami-Havana" and "My Story, My Goal" ensured that the stories that were eventually told were not randomly chosen, but were instead based on a contextual, critical understanding of the subject matter and the larger issues that were going to be addressed. The result was a collection of personal, intimate stories, creatively told, but grounded firmly in reality and speaking to matters of global concern. The framework for teaching multimedia that has been presented here, drawing on the tenets of media literacy and digital fluencies, while ambitious, holds the potential to address both of these challenges: gaining new perspectives on the directions in which the field of journalism is headed and embracing the potentials for incorporating digital technologies more completely into the classroom, while at the same time finding new ways for envisioning the future of higher education and the development of students that are critically engaged, collaborative and global citizens.

REFERENCES

Banks, J. A. (ed.). (2007). *Diversity and Citizenship Education.* Indianapolis, IN: Jossey Bass.

Barnett, S. (2002). "Will a Crisis in Journalism Provoke a Crisis in Democracy?" *The Political Quarterly* 73, 400-408.

Bird, E. (2009). "The Future of Journalism in the Digital Environment." *Journalism* 10, 293-295.

Buckingham, D. (2003). *Media Education: Literacy, Learning and Contemporary Culture.* Cambridge: UK, Polity Press.

Burn, A., & Durran, J. (2006). "Digital Anatomies: Analysis as Production in Media Education." In D. Buckingham and R. Willett (eds.). *Digital Generations: Children, Young People and New Media,* 273–293. Mahwah, NJ: Lawrence Erlbaum.

Deuze, M. (2004). "What is Multimedia Journalism?" *Journalism Studies* 5, 139-152.

Deuze, M. (2005). "What Is Journalism?: Professional Identity and Ideology of Journalists Reconsidered." *Journalism* 6, 442-464.

Downie, L., Jr., & Schudson, M. (2009). "The Reconstruction of American Journalism." *Columbia Journalism Review* (October 19). Retrieved on 20 December 2010 from http://www.cjr.org/reconstruction/the_reconstruction_of_american.php

Dupagne, M., & Garrison, B. (2006). "The Meaning and Influence of Convergence: A Qualitative Case Study of Newsroom Work at the Tampa News Center." *Journalism Students* 7, 237-255.

Gibson, K. L., Rimmington, G. M., & Landwehr-Brown, M. (2008). "Developing Global Awareness and Responsible World Citizenship With Global Learning." *Roeper Review,* 30, 11-23.

Hobbes, R. (1998). "Building Citizenship Skills through Media Literacy Education." In M. Salvador & P.M. Sias (eds.). *The Public Voice in a Democracy at Risk,* 57-76. Westport, CT: Praeger Publishers.

Howland, K. (2006). "Science, Diversity, and Global Learning." *American Association of Colleges and Universities: Diversity Digest* 9, 1.

Huang, E. (2009). "Teaching Button-Pushing versus Teaching Thinking: The State of New Media Education in US Universities." *Convergence: The International Journal of Research into New Media Technologies* 15, 233-247.

Jones-Kavalier, B. R., & Flannigan, S. L. (2006). "Connecting the Digital Dots: Literacy of the 21st Century." *Educause Quarterly* 2, 8-10.

O'Brien, D., & Scharber, C. (2008). "Digital Literacies Go to School: Potholes and Possibilities." *Journal of Adolescent & Adult Literacy* 52, 66-68.

Palfrey, J., & Glasser, U. (2008). *Born Digital: Understanding the First Generation of Digital Natives*, New York: Basic Books.

Presley, P. (2010). "Sunrise, Sunset." *Online Quill* 98. Retrieved on 20 December 2010 from: https://www.spj.org/quill_issue.asp?ref=1736

Quill, S. (2005). *Convergent Journalism: The Fundamentals of Multimedia Reporting.* New York: Peter Lang.

Quinn, S. (2005). "Convergence's Fundamental Question." *Journalism Studies* 6, 29-38.

Resnick, M. (2002). "Rethinking Learning in the Digital Age." In G. Kirkman (Ed.), *The Global Information Technology Report.* World Economic Forum, 32-37.

Chapter 7 -
Incorporating In-Depth Research
Methodologies and Digital Competencies with
Media Literacy Pedagogies

JAD MELKI

American University of Beirut, Lebanon

Introduction

The question of including multimedia production skills—referred to here as digital skills—in media and news literacy classes has long been a contentious point. Hobbs (2001) listed it among the "seven great debates" in media literacy. Proponents of digital skills view media literacy as lacking if it excludes teaching students how to "write" media, along with the critical "reading" of media texts. On the other hand, opponents dismiss digital skills as vocational teaching for underachieving students, and that they do not deserve a place within the walls of the university. Historically, the latter view has dominated academe partly aided by the high cost of production equipment, the difficulty of training academics on technical production skills, and the domination of the various communication and media studies departments by professors who lack these skills.

Today, however, the low cost and ease of using production technologies and the ubiquity of digital and social media have supported the former camp. Teaching video production, that once required massive equipment, lab space, budgets, and multiple experts and technicians, today can be conducted single-handedly using a camera-equipped cellphone and a basic laptop. Even the once expensive video editing and graphic design software and hardware have become standard components of the Windows and Apple operating systems or even available online for free.

Still, the question of the relevance and value of digital skills to media literacy remains important. The common arguments supporting this notion state that students cannot become truly critical consumers of media without producing media texts themselves, and that the best way for students to understand the construction of media texts is through engaging in media text construction activities (Livingstone, 2004). This chapter supports the essentialness of digital skills in media literacy for three additional reasons.

First, digital skills have become essential professional and life skills. An increasing number of industries view digital and social media skills as crucial when hiring new recruits, and that extends beyond the communications, marketing, and news industries (Schwartzman et al., 2009; Knorr, 2009). In addition, digital skills or what the Economist's Digital Economy Ranking (2010) calls web-literacy have become an important indicator of the competitiveness and ability of countries to use Internet and communication technologies (ICT) to advance their economic and social benefit. Moreover, the recent upheavals in the Arab world testify to the role of digital and social media in affecting social and political change and empowering individuals.

Second, the traditional lines that separate "reading" from "writing" and even using from non-using media continue to blur. With the continued convergence of various media tools and the increasing submergence of our daily communication practices in mediated-realities, "we must look at digital literacy as another realm within which to apply elements of critical thinking" (Jones-Kavalier & Falannigan, 2006). As digital technologies become more sophisticated, portable, and ubiquitous, digital tools render the notion of media use as a separate human activity meaningless. The ubiquity of media messages and technologies continue to submerge most social strata to the point that it has become difficult to perceive a time of the day when people are not consuming or producing media texts, sometimes simultaneously. In fact, recent studies of media consumption and production habits had to take into consideration simultaneous multi-media use (Melki, 2010; Rideout, Foehr, & Roberts, 2010). Similarly, the formerly separate tools used to read and produce text, imagery, sound, and moving pictures have become largely one and the same, or at least activities that can be done concurrently.

Third, and most importantly, teaching digital skills not only helps media literacy students become more critical consumers of information but also adept—and critical—producers of information and empowered global citizens able to engage in discussions and organize in networks that shape their societies and improve their individual statuses. And here also lies the critical link between digital and research skills.

Both digital and research skills convert students from consumers to producers of information. Of course, digital and research skills may fall on opposite ends of a spectrum of information quality with research reaching deep into the high quality side of the spectrum. But the increased availability of Web tools relevant to scientific research activities—for data gathering, processing, sharing, presenting, or disseminating—has created a symbiotic relationship between research and digital skills, as this chapter will demonstrate.

Technologies that facilitate conducting research have come a long way since the initial calls for more integration of research into social sciences courses (Makhaim, 1991; Johnson & Steward, 1997). Incorporating research projects into media literacy classes offers multiple advantages. It allows students to experience firsthand and hopefully appreciate the role of research in generating reliable and valid information that today is often published online alongside news and entertainment. This in turn helps students more effectively distinguish different types of information and become critical consumers of research- and science-generated claims. It also creates a partnership between students and teachers and allows students to be part of generating high quality research in media literacy, particularly through rigorous projects that may be published in academic journals. This offers an added advantage to faculty who often struggle to balance teaching and research duties, through efficiently doing both and gaining the help and insight of their students.

While this all sounds good, still the question of effective integration of research and digital skills into critical media literacy instruction offers many challenges and requires more than just a list of benefits and facilitating web tools. The "fears that media production can easily be taught as a decontextualized set of tasks that teach students a narrow set of skills" (Hobbs, 2001, 20) remain valid and could extend into incorporating research skills. Therefore, this chapter aims to offer a practical guide and some examples and suggested tools for effectively integrating research and digital skills within the context and framework of critical media literacy teaching.

The chapter starts with two sections outlining the critical elements needed to ensure effectively integrating digital skills and research projects, respectively. After that, it discusses some cases and examples where digital and/or research skills were integrated into media literacy coursework.

Integrating Critical Analysis with Digital Competencies

My media literacy classes require students to complete a series of assignments throughout the semester, divided between Media Literacy Critical Analysis papers (ML papers) and Digital Skills assignments (DS assignments). The former asks students to critique a media text—advertisements, news articles, images, etc.—linked to a topic covered in class, while the latter entails the production or manipulation of a media text—blog, podcast, video, slideshow, etc.—based on one or two basic training sessions where students learn hands-on how to use the software.

Key in constructing these assignments is to create a clear and explicit link between the ML paper and its corresponding DS assignment, and, of course, a link between both assignments and the course modules. The ways to link ML and DS assignments are numerous and the creative task is often fulfilling. The cases presented below offer some ideas, but teachers can practically build a DS assignment around any ML paper that deals with a media text, process, or audience. At the very minimum, any ML paper can be blogged, audio recorded and podcasted, or converted into a slide show, but that is uninspiring and certainly not what we mean by integrating the critical and the digital.

In addition to linking assignments together, teachers should keep in mind that media literacy classes aim to instill in students a sense of independent learning and initiative. Just as we expect students to independently apply critical thinking skills across any media texts, not only those covered in class, teachers should approach digital skills in the same fashion. They should teach students the logic that governs the use of digital technology. And just as we use examples, demonstrations, and exercises in teaching the critical reading of media texts, we also need to give students basic hands-on instructions for working digital tools. This may complicate matters, as multiple methods are required to ensure that students with different learning methods can understand and utilize digital applications effectively. More importantly, most media literacy classes will bring together students with varying digital competencies, and the challenge is to offer exercises and instructions that engage both students who possess little or no digital skills and students whose skills exceed that of their teacher.

Four guidelines in teaching digital skills may address these issues: Visual demonstrations, one-on-one guidance, non-linear exercises, and scaffolding. For most students, reading a long list of detailed instructions is tedious and frustrating, and they can work faster after watching a short demonstration accompanied by some tips and warnings about common mistakes. Still, other students require more one-on-one guidance to avoid frustration when they get stuck in a certain task and fall behind the rest of the class. Quite often, this only requires quick and basic feedback to nudge the stuck students forward.

But it can also be time consuming, and students may find themselves frustrated waiting for the teacher to reach them, especially in large classes.

Here, offering a non-linear exercise where students can skip to a different task while they await the teacher's help will save time and lower frustration levels. In addition, effective digital assignments utilize flexible software and offer extra optional tasks, "allowing independent exploration for more advanced students, alongside instruction and guidance" (Willett, 2007, 175). Moreover, these advanced students can offer great help in class. Whenever a student finishes her exercise quickly or offers tangible evidence that her skills exceed the basic exercise requirements, I recruit her to help other students during the lab exercise, but also instruct her not to give too much help to the point that she is doing the work for the other students.

Finally, choosing familiar applications and exercises that utilize and build upon students' existing knowledge and skills helps encourage students to explore and learn independently. Scholars refer to this approach as scaffolding (Willett, 2007) or putting in place structures that offer students a familiar starting point to build upon their existing knowledge. These structures are then gradually withdrawn as students master and internalize the new skills and move progressively into unfamiliar areas with confidence. Scaffolding may also include one-on-one guidance, and in the broad scheme of a media literacy course, it guides the manner in which digital skills lesson are lined up, with later applications, for instance video editing, containing skills and knowledge garnered from earlier applications, such as audio editing.

So, a typical digital session starts with a 15-minute lecture about, for instance, digital audio formats and the concept of push technology used by podcasting applications. Then, I deliver a short demo on how to use a digital audio recording application, such as Audacity, to record, edit, and export an audio file. Immediately after that, students follow a written exercise that guides them step-by-step through a specific task, such as creating a 30-second audio interview with one sound bite. The detailed written instructions are left online for students to refer to when they later work alone on their future digital skills assignment, but not for their final project. However, the written instructions are only a basic map for students to follow. So, while students work on their exercise, I shuffle between quick one-on-one answers and repeating parts of the demonstrations. This ensures that students do not miss certain minor procedures not captured in the written instructions.

Furthermore, to insure that students become self sufficient in learning new technologies, and to expose other students to what is out there, one class assignment asks students to search for a "Web 2.0" tool, research it, present it in class, and blog about it. The students have to devise a step-by-step "how-to" guide for their tool and demonstrate evidence of how they used it. In addition, students offer examples of how the Web 2.0 tool can be used in different contexts to empower individuals and communities. Quite often, I learn about new online tools from these presentations, and sometimes a student's presentation is incorporated into the next semester's digital skills repertoire.

Integrating Research: Lessons from the Natural Sciences

Research projects incorporated into media literacy classes may vary from simple informal components of a project meant mainly as exercises or pilot studies to full-fledged rigorous research projects that may span several courses. Regardless the scope, teachers must devise a specific purpose for the project that clearly fits the course objectives and learning outcomes. The purpose should clarify the role of the student in the project and how they may benefit from it. Will the project incorporate the students as researcher or research assistants? Or will it engage them as research participants? The latter is sometimes harder to justify, but as our examples below about media habits and media dependency show, students benefitted immensely from the experience and the reflexive writing component.

For a while, the natural sciences have been able to incorporate research in their courses and labs more effectively than the social sciences and humanities. So, it makes sense to learn some lessons from them. In a workshop on integrating research into geoscience courses, Beane (2005) offered six tips on how to effectively incorporate research in courses:

First, "pose a hypothesis" to invite students to explore further applications of your lessons (Beane, 2005). Take for instance case 6 below about the concentration of media ownership (Who Owns the Media in Your Country?). A teacher can set a research question focusing on ownership concentration in a specific subset of the industry, say Internet Service Providers, or she can hypothesize about general media ownership trends in a different country, for instance mass media ownership trends in Lebanon. The question can then be followed by an assignment asking students to gather information about media ownership and assess the extent of ownership concentration in that industry (see case #6).

Second, ask students to "write a proposal" to encourage them to think about their own research questions and perhaps follow through in answering them during the semester (Beane, 2005). These proposals can also be group projects or even subsets of one broad media literacy topic for the whole class, where each group of students works on a specific angle of the topic.

Third, require a "background research" or literature review that helps students become more effective at accessing scientific research (Beane, 2005), distinguishing it from other types of information, understanding its structure and content, and develop critical skills for reading scientific research. This also helps the teacher gather literature on a topic she may be researching. This kind of research exercise better fits advanced media literacy courses and requires some lecturing about how to access, evaluate, and analyze scientific literature.

Fourth, "data collection" could happen in different forms and various scopes (Beane, 2005), but key to successful data gathering is good research design, effective communication between the group members, and efficient data gathering tools. Comparative content analysis of media texts—qualitative and/or quantitative—offers an excellent method for use in media literacy classes. As presented in case 1 below (Global News and Content Analyzing the Beijing Olympics), content analysis projects could have students from one semester analyze a sample of media content—advertisements, news articles, images, etc.—or the project could span several semesters. The latter offers the advantage of comparing findings over time. Other data gathering methods could include surveying media audiences, interviewing media managers, using ethnographic research to observe certain media-related activities, etc.

Next, "data analysis" may rotate around data sets gathered by professors or students in prior classes or may utilize data sets readily available online or in the library (Beane, 2005). In a typical media literacy course, however, data analysis should remain at a basic level. Although the teacher at a future stage can clean up and analyze the gathered data in a more sophisticated manner, this may take substantial time and require advanced skills not necessarily available among students of a typical media literacy course. Still, students have plenty of data analysis options even at the most basic levels. As mentioned below, with the innovative use of online tools, such as online surveying applications, the process could be quite simple and almost instantaneous. Cases 1, 2, and 5 below offer many examples of such tools and how to deploy them.

Finally, having students present their research findings offers various advantages to a media literacy class. Presentations, in general, are essential and often considered a cornerstone of media literacy teaching (Mihailidis, 2009). They not only offer students the opportunity to enhance their public speaking skills but also allow the whole class to reflect on the research results and enforce and enrich the media literacy lessons. This is particularly true in projects where different students work on different components of the same topic, and the presentations reveal the big picture of the findings, such as the case of the tobacco billboards discussed below. But presentations need not be only oral. Digital production skills students learn in the same media literacy class offer them unlimited venues to deliver and share the findings with the world, and to demonstrate their research skills and digital prowess.

Regardless of the ways a teacher incorporates research into a media literacy course, a plethora of digital tools, many freely available online, can help ensure the success of the project. These tools can help in at least three areas: Effective team communication and collaboration, efficient data gathering, and simple data analysis and presentation.

Incorporating research in a class, especially collaborative research, does not differ from leading a research team. The most important matter entails keeping everyone on the same page and well aware of what they should do, in what manner, and when. A content analysis coder, for instance, needs to know specifically which content she will analyze, how to access the codebook, where to submit the data, and by what deadline. Imagine having 30 students conducting 30 separate codings over three weeks, and you can easily see how the system could breakdown in an instant, not to mention create a frenzy of emails from students confused about where to go and what to do. A blog or simple Google web site can create a common document for everyone to follow updates and directions, but individual students or groups may have different steps to follow in addition to the general directions for the whole class. The Google Docs' spreadsheet application offers tremendous help. If each of your students is allocated an ID number, he can browse to the online spreadsheet, locate his ID, and figure out the tasks at hand. Figure 1 below is a spreadsheet used for a collaborative content analysis of international newspapers' coverage of the Beijing Olympics (details of this class project below). Each row contains an ID for the coded material (Content ID), the country, language and name of the newspaper, the ID numbers of two students (Coder ID), the story ID number, and the URL for the newspaper article to be analyzed. The purpose of having two coders for each article is to measure coding reliability.

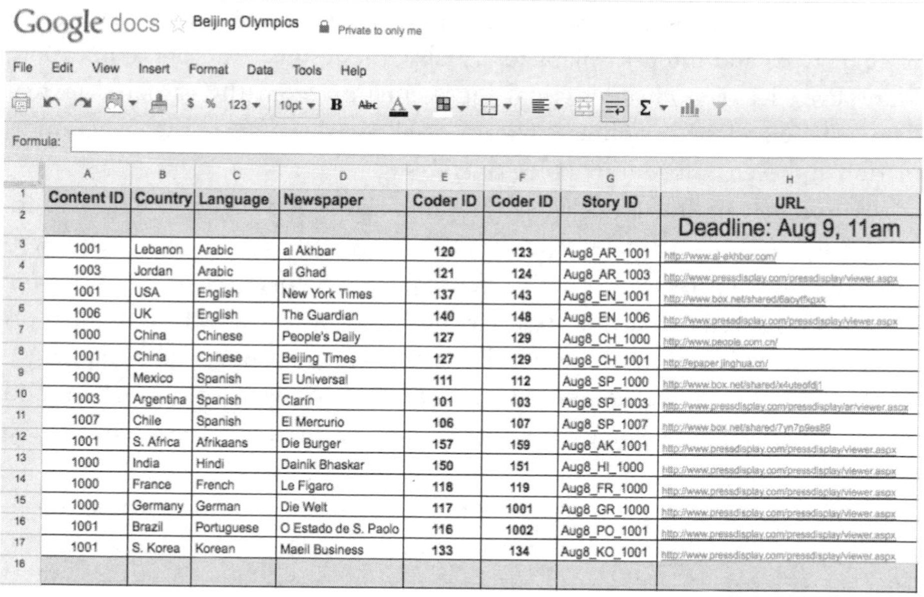

Content ID	Country	Language	Newspaper	Coder ID	Coder ID	Story ID	URL
							Deadline: Aug 9, 11am
1001	Lebanon	Arabic	al Akhbar	120	123	Aug8_AR_1001	http://www.al-akhbar.com/
1003	Jordan	Arabic	al Ghad	121	124	Aug8_AR_1003	http://www.pressdisplay.com/pressdisplay/viewer.aspx
1001	USA	English	New York Times	137	143	Aug8_EN_1001	http://www.box.net/shared/6aoytfkgxk
1006	UK	English	The Guardian	140	148	Aug8_EN_1006	http://www.pressdisplay.com/pressdisplay/viewer.aspx
1000	China	Chinese	People's Daily	127	129	Aug8_CH_1000	http://www.people.com.cn/
1001	China	Chinese	Beijing Times	127	129	Aug8_CH_1001	http://epaper.jinghua.cn/
1000	Mexico	Spanish	El Universal	111	112	Aug8_SP_1000	http://www.box.net/shared/x4uteofdj1
1003	Argentina	Spanish	Clarín	101	103	Aug8_SP_1003	http://www.pressdisplay.com/pressdisplay/ar/viewer.aspx
1007	Chile	Spanish	El Mercurio	106	107	Aug8_SP_1007	http://www.box.net/shared/7yn7p9es89
1001	S. Africa	Afrikaans	Die Burger	157	159	Aug8_AK_1001	http://www.pressdisplay.com/pressdisplay/viewer.aspx
1000	India	Hindi	Dainik Bhaskar	150	151	Aug8_HI_1000	http://www.pressdisplay.com/pressdisplay/viewer.aspx
1000	France	French	Le Figaro	118	119	Aug8_FR_1000	http://www.pressdisplay.com/pressdisplay/viewer.aspx
1000	Germany	German	Die Welt	117	1001	Aug8_GR_1000	http://www.pressdisplay.com/pressdisplay/viewer.aspx
1001	Brazil	Portuguese	O Estado de S. Paolo	116	1002	Aug8_PO_1001	http://www.pressdisplay.com/pressdisplay/viewer.aspx
1001	S. Korea	Korean	Maeil Business	133	134	Aug8_KO_1001	http://www.pressdisplay.com/pressdisplay/viewer.aspx

Figure 7.1: Google Docs Spreadsheet for Organizing Researchers. Courtesy of Author

Various other Google tools can also be used for collaboration. For example, the Document and Presentation tools offer effective methods for multiple people to edit the results of the research and build a slideshow presentation out of the results.

The second most important matter is finding an efficient data gathering and processing tool. Online survey applications–many of which offer free basic access–can save you and your students substantial time and effort, and these applications are not limited to conducting surveys. Any survey application can be easily converted into a codebook instrument for a content analysis project, or even a data collection tool for qualitative interviews and paper-surveys. In addition to offering a simple mechanism for data gathering, these same tools also allow for a centralized storage area for the collected data and eliminate the need for manual data-entry and collation. They also allow sharing information, monitoring work, and analyzing data.

Many online data gathering tools offer simple displays of frequency tables, cross-tabulations, and filtering of certain variables. Some also help in qualitative data analysis and allow for displaying simple charts. But if you already have the information sorted in a database, you can also take advantage of the many data visualization and mapping tools available online. One application,

Google Fusion Tables, lets researchers upload large datasets and use a variety of chart types and intensity maps to visualize the results. Another application, Many Eyes, lets you create a visual representation from the gathered textual data. These and other online applications allow you to share the visual information and even embed it in a blog or web site.

Before delving into cases where research skills—and digital skills—were incorporated into media literacy classes, teachers should realize that incorporating research in any course requires extensive prior preparation and monitoring during the semester. But if a project is designed properly and the right tools applied, the extra time and effort will often pay off through student appreciation, efficiently conducted research, effectively improving your teaching and course material, and in some cases producing a research publication out of the work.

The rest of this chapter discusses a few cases and examples where digital and/or research skills were incorporated into media literacy courses.

Case Studies in Research/Digital Competencies in Media Literacy Education

1. Global News and Content Analyzing the Beijing Olympics

In the past few decades, comparative content analysis garnered the attention of a growing number of media scholars trying to understand the incongruous news coverage of international events by different media systems (e.g., Siebert, Peterson, & Schramm, 1963; Norris, Kern, & Just, 2003; D'Angelo & Kuypers, 2009). Recent technological developments that brought once distant and inaccessible media content to the fingertips of almost anyone around the world, accompanied by advancements in research and computing tools, made content analysis more efficient—and in demand. However, up until very recently, much of this research concentrated on English texts and major Western media outlets. Despite all the advancements in this methodology, major challenges still abound. These include dealing with multiple languages and news formats, storing, sampling, and organizing the enormous amount of news sources and analyzable data, and training coders of diverse backgrounds and cultures. But a media literacy course with enthusiastic and talented students and the right mix of tools can overcome these obstacles.

The 2008 Salzburg Academy on Media and Global Change,[1] a summer program centered around media literacy education and research from a global perspective, offered an excellent venue for conducting a global content analysis of news while simultaneously teaching students news literacy. The academy that brings together each summer 60 students and 12 media faculty from some 20 nationalities offered an exceptional pool of researchers, linguistic variety, and cultural diversity. Coinciding with the academy was the 2008 Beijing Olympics that offered an ideal research topic, given its global scope and its potential to attract enormous news coverage.

The first challenge was to agree on a scope and purpose for the research project. While most content analysis research today focuses on a narrow aspect of the analyzed text and builds upon framing theory and other related theoretical constructs, our aim was different. The purpose of our project was to capture the multiple influences that may have shaped the Olympics coverage and to offer an analysis that reflects the richness and multifaceted concepts of media literacy theory. The faculty members first agreed on the areas the analysis will cover: The prominence of the coverage, the dominant sources, the level of ethnocentrism, the extent of sexualization of athletes in the images, the tone of the coverage, the tone towards China, the level of objectivity, and the framing of the topics, athletes, and countries. For efficiency and due to limited time, the analysis focused on newspaper front pages only.

Based on these priorities, a faculty member and a group of graduate students, representing each of the languages used, constructed a codebook using SurveyMonkey, an online surveying tool. After multiple pre-tests and revisions of the codebook, the faculty members agreed on a final version, and each faculty member was appointed to supervise a group of student-coders. All students underwent multiple training sessions until an acceptable inter-coder reliability rate was achieved. Each of the first three training sessions used the same English news article. Then a fourth session used articles from the various languages.

The second challenge was sampling the content and organizing the coding process. Each day, the team of graduate students downloaded the designated newspaper front pages, mainly from Newseum's Front Pages web site, uploaded them to Box.net, an online file-sharing service, and organized their links in a Google Docs spreadsheet that each student accessed for her daily coding assignments (see the Google Docs example above for details). Short daily briefings addressed any common concerns and offered feedback to students based on monitoring the analysis and the coding reliability. To effectively obtain inter-coder reliability coefficients, at least two students coded each newspaper front page, and the consistency of their answers was compared using Excel. Within seven days, 59 students analyzed 484 front pages of 68 newspapers from 29 countries in 10 languages: Afrikaans, Arabic, Chinese, English, French, German, Hindi, Korean, Portuguese, and Spanish. All this and they achieved an inter-coder reliability rate that exceeded 90%.

During the last day of the academy, the research director presented the findings to a captivated audience of international student-coders and media literacy professors, who saw the fruits of their hard work. Many of the findings were not expected and both students and faculty were surprised. For example, the group of students and professors from China were quite anxious throughout the academy to know how global media covered their country. Their relief was sensed throughout the presentation hall when the results revealed that the press predominantly gave China favorable coverage. Other interesting moments corresponded to the revelation that the coverage differed was highly ethnocentric and male-centered, where female athletes received half as much coverage as their male counterparts, with the exception of Africa where the press gave more space to female athletes.

The whole study was then published online and appeared that year as one of the main tangible outcomes of the 2008 Salzburg Academy (Melki, Moeller, Mihailidis, & Fromm, 2008). When the Chinese group returned home, they too translated and published the study that received substantial press coverage locally. More importantly, the project engaged a global group of media literacy students and professors in an exercise that not only made them more critical readers of media texts but also gave them the chance to produce high quality knowledge and earn research skills in an exciting and efficient manner.

This same approach was used in regular semester courses to content analyze international news coverage of war and political violence in the Arab world. In this case the class instructor was responsible for sampling the data and setting up the grid well before the class started. The project took four semesters and eight courses to complete the coding of almost 1,000 TV news reports.

2. Media Dependency, Media Habits, and Going 24 Hours Unplugged

Not all research projects should be so extensive. In one media literacy course Moeller (2010) asked her University of Maryland students to "go unplugged for 24 hours." This meant students used no modern media tools, including cellphones, emails, computers, iPods, TV, radio, etc. for a full day. The exercise focused on the concept of media dependency (Ball-Rokeach & DeFleur,

1976). Students reported back their experiences through blogs, and Moeller analyzed their reflexive essays unveiling simultaneously fascinating and disturbing themes. Among the findings, students used "literal terms of addiction" to describe their dependency; they found the separation from their media tools unbearable and frustrating; and many expressed feelings of loneliness, seclusion, and anxiety and felt that detaching themselves from their media tools was akin to separating themselves from their friends and family. In addition, the students learned firsthand how media-dependent they were and the media-centered habits they had unconsciously developed over time. They realized how much they were missing by being constantly "plugged," and how these habits affect them emotionally and sometimes even physically. The simple study attracted immediate media attention and generated a wealth of data—over 110,000 words of reflexive critical writing, according to Moeller (2010).

The following year, Moeller (2011) invited media literacy professors from 12 universities—from Argentina, Chile, China, Lebanon, Mexico, Slovakia, Uganda, the UK, and the USA—to replicate this same exercise. The outcome was a global research project that engaged some 1,000 media literacy students in an experiential media-dependency lesson. The online tools Moeller and her team used were key to the success of this global project. To keep track of the demographics, especially which countries and universities participated so the data can be compared later, the team asked participants to fill a short questionnaire using the online surveying tool *SurveyMonkey*. During the pre-testing phase of the survey, the questionnaire was limited to demographic questions, but it later incorporated some quantitative measures of media consumption habits, which enriched the data and added a new dimension to the analysis.

The idea of adding media habits questions came from a similar research project that focused on media consumption and production habits of Arab youth and surveyed students from Jordan, Lebanon, and the UAE (Melki, 2010). Before deploying this latter study, it was piloted in a media literacy class where students offered insight about their media habits. The piloting engaged filling the survey questionnaire and offering feedback about questions and answer options, particularly the scales that measured frequency and level of media use. The students also helped in distributing the questionnaire for piloting and reported back critical information about how long it took to administer it and what questions participants found difficult to answer. Students indirectly learned the nuances of surveying and saw how prevalent certain media habits were. The process also generated excellent material for discussion in class and helped the students become more aware of their own media habits.

3. Critiquing Magazine Covers and Photo-Editing Skills

Most media literacy courses teach about the persuasive power of images and the skewed representation of reality manifested through the ubiquitous practice of photo manipulation. Adding a hands-on photo-editing module to the course can naturally enrich the students' understanding of this topic while simultaneously teaching them a new production skill.

The first time I tried to incorporate basic PhotoShop lessons in the course, the hands-on sessions happened to coincide with lessons about ownership trends and the influence of business on media. The two sessions that week had little to do with each other and might as well been from two separate courses. The second time around, the photo editing exercises were aligned with lectures on the power of photography in news. The students were more easily able to draw connections between the lectures and readings and the photo editing exercises, and the discussions spilled from one module into the other. Still, something was missing. The photo editing assignment had nothing to do with the critical essay on photography students had to write. So, I tied them together, the third time around. The critical paper asked students to find a magazine cover and write a paper critiquing the choice of an image or photo illustration. It also asked the students to play the role of editor and come up with three alternative photo ideas to represent the same story.

The subsequent week in the digital skills assignment, students implemented the alternative image ideas they devised in the critical assignment by erasing or altering the original image from the magazine cover and adding instead one of the alternative photos or illustrations they suggested.

The assignment combo generated positive feedback, but more importantly students learned "hands-on" about the ethical and technical challenges and limitations—and sometimes lack of good judgment—journalists faced in selecting and editing images. The students also learned about Creative Commons Licensing and how to legally and inexpensively acquire high quality images to use in their assignments, and with proper citation and referencing.

Another media literacy course aligned the photo-editing module with lessons about advertising and body image. In a critical essay, students analyzed a group of hyper-sexualized banner or magazine advertisements. In the digital skills assignment, they created alternative ads for the same products.

Some of the most interesting assignments analyzed portrayal of women in various ads and demonstrated how the same products could present women in a less sexually objectifying manner while simultaneously offering a more inclusive picture of various body types, ages, and races.

4. News Constructions, Editing, Podcasting, and Vodcasting

Teaching media literacy students how journalists and news operations frame news stories poses many difficulties. Primarily, students jump into conspiracy theories about the intentions of the news media and often miss the important media literacy concept of news as a constructed message influenced by various political, commercial, institutional, cultural, and technological factors.

One way to help media literacy students understand the pressures, biases, and limitations of news production is to have students themselves construct and frame news stories. And if you are planning to teach students basic video production, why not combine the exercises? Because shooting, scripting, and editing video could easily occupy a full course or more, some limitations need to be imposed on video lessons in a media literacy course. And to ensure that the practical work does not become decontextualized from the critical media literacy lessons, teachers should make sure they alternative between lessons and discussions about news framing, on one hand, and practical work on audio and video production, on the other hand.

The best place to start the skills classes is with simple audio editing. Students may use a freeware, such as Audacity, to create a 45-second audio report with two 10-second sound bites. For efficiency, I usually supply them with a pre-recorded interview with roughly 10 sound bites, although advanced students may record their own interviews. The assignment can ask students to write three sentences: An introduction to the report that also introduces the first sound bite; a bridge joining the two sound bites; and a conclusion wrapping up the report. After learning how to edit the recorded sentences and sound bites into one coherent report, they can upload their audio segments to a free podcasting service, such as Podbean or Podomatic, and embed it in their blog, while they simultaneously learn about RSS feeds. The class then can listen together to all the finalized reports and observe how different students constructed different version of the same news story through the selection of sound bites and choice of words in their narrations. Students are often surprised about the divergence in their productions and express better understanding of news framing—and pride their 45-second achievement.

After students become comfortable with audio editing, the digital skills that naturally follow—and provide for scaffolding from the previous skills—are video editing. The same exercise above can be easily replicated to fit a TV news report, where most of the components, especially the interview and the video footage, are supplied beforehand. The students can upload their work to YouTube and embed it in their blog or web site.

Multiple other media literacy lessons can be attached to the video editing modules. For example, lessons about propaganda techniques used in documentaries are much easier to comprehend by someone familiar with the nuances of video editing. Students can experiment with the propaganda tactics of "pacing and distraction" or "association" (Rhoads, 2004). They can experience how the emotive level in their video or audio segments increase with the use of music.

Although video editing is more complicated than audio editing, a well-prepared exercise broken down into small steps often flows smoothly without a glitch. Building digital skills chronologically also helps make the process efficient. It makes little sense to start with video editing and then move to audio and photo editing, as skills from the latter build towards the former. Over the years, I have been able to cover basic video editing skills in as little as two 90-minute sessions. I'm always surprised about the ability of students to create videos after so little training. Many students end up submitting a Vodcast or Podcast for their final project.[2]

5. Advertising, Tobacco Billboards, and Online Mapping

Many media literacy students enjoy analyzing advertisements, and even more so photo editing or video editing an existing commercial to make it more inclusive or less sexually objectifying or simply less stereotypical. But one media literacy class took the advertising lessons and exercises to a new level that mixed critical analysis, digital skills, and research altogether.

Each group of two students was tasked with locating 15 tobacco billboard advertisements, which clutter the streets of Beirut. The students were responsible for snapping a digital photo of each billboard and writing a short report about the location, content, and the persuasion techniques used in each ad. When all the students presented their findings, the class suddenly realized that many of the billboards were within striking proximity of a school or university campus and most billboards used advertising techniques particularly effective with teenagers and young adults.

The next semester, we attempted to repeat the same exercise and added a Web 2.0 component. In addition to the preceding tasks, students also had to upload the images and brief analyses to a Google map and highlight the schools and universities within the vicinity. However, the project coincided with the proposal of new Lebanese media regulations that limited tobacco advertising, and the preemptive reaction of the tobacco industry left very few ads to report.

The next time around, we plan to expand this project further and include a crowdsourcing component. Using Ushahidi's new Crowdmap application and a few common social media tools, we hope the media literacy students will be able to solicit the help of their friends to locate billboard ads that violate the law, both in their content and location.

6. Who Owns the Media in Your Country?

Teaching the political economy of media outside the USA and Europe poses several challenges, most important of which is the lack of literature on the matter in many developing countries. The lack of transparency and ineffective laws and institutions that allow access to public information offer formidable obstacles. But how can we expect students to become media literate if they don't see the ownership patterns of media institutions in their countries?

This problem offers a research opportunity with two purposes. Gathering reliable media ownership records enhances students' abilities to access information, while simultaneously unveiling the real movers and shakers of the media industry in their country. Of course, this exercise may differ radically from one country to another, but Lebanon offers an excellent example due to the unavailability of such information beyond scattered and mostly inaccurate news reports and due to the immense difficulties one faces in acquiring basic ownership documentation.

The first time we attempted this project, we hit a solid wall. Students had three weeks to conduct preliminary research about a broadcast media institution of their choice, and for extra credit they should acquire official ownership documents of the media institution in question. None of the visited companies divulged ownership information, and some even got deeply offended by the request and abruptly ended the interviews. All these companies also denied that political figures or parties owned their companies.

Next, despite multiple promises at first, the Lebanese National Council for Audio-Visual Media—the official body that recommends the licensing of TV and radio stations and thus possesses all the public ownership records—supplied us with no documents. Several other venues proved to be dead ends, and by then most students gave up.

However, an intellectual property lawyer recommended visiting the Trade Registry Office, where all commercial companies deposited their public records and updated them annually. This lead proved fruitful but not without major obstacles. The civil servants at the registry office refused to respond to college students; so, some of the students dressed up professionally and posed as legal researchers. The trick worked, but the trade registry officials could only access the records using the company's trade registration number. That led the students to another government office that can search its computer databases for the registration number based on the name of the company. The students quickly learned that almost all companies' registered names differed from their public names. This brought them back to the National Council for Audio-Visual Media who at least was willing to disclose the official name of the companies under research.

In the end, only three out of 25 students succeeded in acquiring official ownership documents, and the three of them quickly found out that prominent politicians and members of certain political parties were major stockholders in all of the three companies researched. They also found that the few news reports about this matter were entirely inaccurate. The students learned first-hand about accessing reliable ownership information, the difficulties and frustrations in doing so, and the structures of power revealed when such information was divulged. The experience also offered future students a roadmap for accessing this information, and indeed over two years students were able to accumulate a formidable amount of official records about media ownership that will be the basis for a future study on the political economy of media in Lebanon.

Conclusion: Integrating Research and Active Production for Media Literate Outcomes

By the end of a media literacy course, most students develop keen analytical abilities to read media texts, and become adept at using multiple digital media tools. So, for a final project in media literacy, it would make more sense to adopt the mode of "how can we help others become media literate?" instead of asking students to analyze more media texts. This also fits perfectly with the incorporation of digital and research components, and students become active contributors to the media literacy body of knowledge.

In my final projects, students have to include five components: A narrowly defined media literacy topic; a background study and literature review of the media literacy topic addressed; original data or analyses that uses a clearly de-fined research method or media literacy analytical approach; a "tool kit" that contains a list of questions or analytical tools that pertain specifically to the study question and that offers the public an analytical instrument to critically analyze the media content in question; and finally a digital component that publishes all the above in a manner easily accessible to the public.

Some of the best final projects submitted include a podcast on the wed-ding industry in the USA and how it is indirectly promoted through movies and TV programs, a web site focused on political propaganda in Lebanese elections, a video about MTV's homogenizing effect on Arab music videos, to name a few. Many of these projects get posted on the American University of Beirut's dedicated YouTube channel (AUB Media Studies, 2010), and some are used in subsequent classes.

While the task of incorporating digital and research skills in media literacy courses can be challenging and requires extensive preparation and follow up during the semester, the outcomes are worth the extra effort and time, both to the students and the teachers, and sometimes to media literacy research and the general public. Effectively integrating digital and research skills into media and news literacy teaching builds on the critical reading skills traditional me-dia literacy classes produce, and helps students transition from media con-sumer to adept and critical producers of information and knowledge and empowered global citizens engaged in important discussions and able to or-ganize in networks better positioned to shape societies and regions and en-hance the statuses of marginalized individuals and disenfranchised groups.

RESEARCH & DIGITAL COMPETENCIES

REFERENCES

AUB Media Studies Program. (2011). *AUB YouTube Channel*. Retrieved on 12 April 2011 from http://www.youtube.com/AUBatLebanon#p/c/640894941A1A2BAA

Ball-Rokeach, S. J., and DeFleur, M. L. (1976). A Dependency Model of Mass Media Effects. *Communication Research*, 3, 3-21.

Beane, R. (2005). "Integrating Research into Geoscience Courses." *The Science Education Resource Center at Carleton College*. Retrieved on 16 April 2011 from http://serc.carleton.edu/NAGTWorkshops/careerprep/teaching/IntegratingResearch.html

D'Angelo, P., and Kuypers, J. A. (Eds.). (2009). *Doing News Framing Analysis: Empirical and Theoretical Perspectives*. New York: Routledge.

Digital Economy Ranking 2010: Beyond Readiness. (2010). Economist Intelligence Unit. *The Economist*, June. Retrieved 16 April 2011from http://www-935.ibm.com/services/us/gbs/bus/html/ibv-digitaleconomy2010.html

Hobbs, R. (1998). "The Seven Great Debates in the Media Literacy Movement." *Journal of Communication*, 48(1), 16-32.

Johnson, M. A., and Steward, G. (1997). "Integrating Research Methods into Substantive Courses: A Class Project to Identify Social Backgrounds of Political Elites." *Teaching Sociology*, 25, 168-175.

Jones-Kavalier, B. R., and Falannigan, S. L. (2006). "Connecting the Digital Dots: Literacy of the 21st Century." *Educause Quarterly*, 2-15.

Knorr, A. (2009). Social-media skills become crucial for job hunters. *The Atlanta Journal-Constitution*, 28 January. Retrieved on 12 April 2010 from http://www.ajc.com/business/social-media-skills-become-102247.html

Livingstone, S. (2004). "Media Literacy and the Challenge of New Information and Communication Technologies." *The Communication Review*, 7, 3-14.

Markham, W. T. (1991). "Research Methods in the Introductory Course: To Be or Not to Be." *Teaching Sociology*, 19, 464-471.

Melki, J. (2010). "Media Habits of MENA Youth: A Three-Country Survey." Youth in the Arab World. The Issam Fares Institute. American University of Beirut, Beirut, Lebanon.

Melki, J., Moeller, S., Mihailidis, P., & Fromm, M. (2008). "Covering the Beijing Olympics." *Salzburg Academy on Media and Global Change*. Retrieved on 16 April 2011 from http://www.icmpa.umd.edu/salzburg/new/lessons/main-findings

Mihailidis, P. (2009). "Beyond Cynicism: Media Education and Civic Learning Outcomes in the University." *International Journal of Media and Learning*, 1(3), 1-13.

Moeller, S. (2010). "24 Hours without Media." *International Center for Media and the Public Agenda, University of Maryland*. Retrieved on 22 April 2011 from http://theworldunplugged.wordpress.com

Moeller, S. (2011). "A Day without Media." *International Center for Media and the Public Agenda, University of Maryland*. Retrieved 22 April 2011 from http://withoutmedia.wordpress.com

Norris, P., Kern, M., & Just, M., (2003). *Framing Terrorism: The News Media, the Government, and the Public*. New York: Routledge.

Rhoads, K. (2004). Propaganda Tactics and Fahrenheit 9/11. *Working Psychology*. Retrieved 15 April 2011 from http://www.workingpsychology.com

Rideout, V. J., Foehr, U. G., & Roberts, D. F. (2010). *Generation M2: Media in the Lives of 8- to 18-Year-Olds*. The Henry J. Kaiser Family Foundation. Retrieved February 15, 2011, from http://www.kff.org/entmedia/mh012010pkg.cfm

Schwartzman, E., Smith, T., Spetner, D., and McDonald, B. (2009). 2009 Digital Readiness Report: Essential Online Public Relations and Marketing Skills. *iPressroom*. Retrieved 15 April 2011 from http://www.ipressroom.com

Siebert, F. S., Peterson, Th., & Schramm, W. (1963). *Four Theories of the Press: The Authoritarian, Libertarian, Social Responsibility and Soviet Communist Concepts of What the Press Should Be and Do*. Urbana: University of Illinois Press.

Willett, R. (2007). "Technology, Pedagogy and Digital Production: A Case Study of Children Learning New Media Skills." *Learning, Media and Technology*, 32(2), 167-181.

Chapter 8 -
Deepening Democracy through
News Literacy: The African Experience

GEORGE W. LUGALAMBI
Kampala, Uganda

Introduction

With many African countries now governed under multi-party systems, and with elections now becoming the rule rather than the exception for the newly democratizing nations, issues are being raised about the substance as opposed to the form of democracy. For many citizens, it is no longer so much a question of whether democracy is feasible in Africa but of how to deepen its roots. This chapter explores the growing relevance of media in answering this question of democratization while building a case for news literacy as a vital instrument for democratic citizenship in Africa.

The analysis is premised on the idea that with democratic participation in Africa increasingly mediated by the media, a news-literate citizenry is no less critical to the quality of democracy than its institutional mechanisms, which tend to receive most of the attention. In driving this point home, the debate about the democratic functions of the media is brought back into focus. Employing the notion of news literacy offers a useful way to situate the role of the media in deepening democracy and in nurturing democratic citizenship in Africa.

Indeed, going by the evidence from recent and ongoing developments, it is unlikely that most of the countries marching towards more liberalized politics today, even if doing so while screaming and kicking, will slide back wholesale into the incorrigible and totally unrestrained dictatorships of the past. That said, it is undeniable that the hybrid character of African democracies is such that in many countries, pro-democratic and anti-democratic forces routinely keep the same company. Obviously, the magnitude of the authoritarian holdovers varies from country to country and mutates from time to time. So the co-existence of these two forces leads some to retain hope about the possibility of positive change; while for others, it is evidence that African countries cannot make a clean, irreversible break with the past. A few countries bear some truth to this skepticism.

In Uganda, elections have not altered the distribution of power in a fundamental way as the government, though thrice returned to power through multi-party elections in the last 15 years, seems to have failed to use its mandate to impose serious and durable restraints on many of those in the governing ranks. Critics point to the endemic corruption in government as evidence of this failure. Transparency International's Corruption Perceptions Index 2010 ranks Uganda as "highly corrupt"[1] and the country's rating has been just about the same over the last five years.[2] The refusal to muster the political will to deal decisively with corruption helps keep the wheels of the patronage system turning. The National Resistance Movement's 25-year grip on power has proven to be a reliable form of political insurance cover for abuse of office.

Kenya arguably went into the elections of December 2007 with a relatively more level playing field among the contesting parties than it had ever experienced since the return to multi-partyism in 1992. In spite of the heated campaigning, the apparent composure in the pre-election period only turned out to be the calm before the storm; for the country imploded thereafter into violence that was as extreme as it was surprising for a country that had been a beacon of stability in the region. At least 1,000 people are estimated to have been killed in the post-election violence, which is now a subject of investigation by the International Criminal Court at The Hague.[3] The power-sharing arrangement that former UN secretary-general Kofi Annan brokered between President Mwai Kibaki and Prime Minister Raila Odinga basically papered over deep-seated concerns about the inability of Africa's political elite to agree to and respect the rules of the game. Predictably, the resultant coalition government has fallen short of the citizens' expectations. A report by Afrobarometer (2010) concluded from a survey carried out in October and November 2008 thus:

> On the one hand, a majority of Kenyans are of the view that creating a coalition government was the best way to resolve the post-2007 election crisis. On the other hand, the coalition government has so far failed to present a united front, a fact that has seen it fail to deliver on the promises of the three parties. The overall verdict from Kenyans is that they are disappointed with the performance of the coalition government thus far. (1)

The deposed leader of Ivory Coast, Laurent Gbagbo, initially seemed to be holding out for a power scheme similar to Kenya's until he opted to fight to retain the disputed presidency at all costs. Gbagbo refused to hand over power to his rival Alassane Outtara. By all credible accounts, Outtara had won the election in November 2010. The intransigent Gbagbo, in brash defiance of public opinion and international norms, dug in and plunged the country into a needless war that claimed the lives of thousands before he was routed and arrested in April 2011.

There are lessons for news literacy in the foregoing examples. The news media, as will be shown subsequently, are often criticized for inducing political apathy in the citizenry. Then again, the reality is that such occurrences and behaviors as witnessed in the African countries cited here do little to instigate or restore people's faith in the political process as a means of consolidating democracy and resolving political disagreements. To understand the potential power and limitations of news literacy in Africa, we need to be aware of both the conditions of the news industry and the unique features along the trail that democracy has followed.

Challenges of the News Industry

Among the myriad of problems troubling the news industry in Africa, as elsewhere, few are arguably more vexing than how to sustain the relevance and vibrancy of the editorial product. The oft-cited source of this predicament is the shift from the old media order under which, according to Meyer (2003), "information was scarce and value was added simply by gathering and delivering it" (12). Under the new media order, however, there are vast supplies of information, which compels the industry to find other means of adding value to news and the editorial output in general. In this new media landscape, "public attention" has become the scarce commodity. To borrow yet again another insight from Meyer (2003):

> This straightforward theory explains most, if not all, of the bizarre behavior of media today. Reality television; the blending of news, entertainment, and advertising; the use of sex and violence in all three; the emphasis on personality over substance – all are calculated to break through the surplus of information and command our attention. (12-13)

Among the critics of the current system of excessive commercialization, Dahlgren (1995) has taken particular issue with the way the audience has been exploited and practically debased in the uninhibited drive for content that aims to appease without provoking.

> Under the commodity logic of the commercial system, the audience becomes the product, delivered to the advertisers, as many critics have pointed out; the programming is – at base – the 'filler' between the advertisements. This logic can help to explain why certain kinds of programmes are aired rather than others: broad appeal and the avoidance of serious controversy are two generally operative criteria. (29)

The two scenarios quoted above, which are emblematic of the wider critique of the current media system, illustrate the difficulty of mobilizing citizens' attention and focusing it on fundamental civic issues. These concerns,
moreover, are not unique to media-saturated societies. The most recent data
available from the International Telecommunication Union (2009) provides
some perspective. Although Africa in general lags behind other regions on
most media access and usage indicators (see Table 8.1 and Table 8.2), emerging trends suggest that media penetration is growing rapidly and you get a very
interesting picture when you disaggregate the trends country by country.

For example, national surveys by AudienceScapes (2010), which were conducted in July 2009, found that 87 percent of Kenyans (sample of 2,000) and
86 percent of Ghanaians (sample of 2,051) had access to a radio at home that
was in working order. Similarly, 71 percent of Kenyans and 72 percent of
Ghanaians reported that they owned a mobile phone. Access to a TV at home
that was in working order was 41 percent in Kenya and 59 percent in Ghana.
The global reach of media technology implies that the dynamics that shape the
performance of the media will, and in some respects already do, have a universal effect on journalistic practices.

Table 8.1 - ICT Uptake in Africa, Developing Countries & the World per 100 Inhabitants, 2008

	World	Developing Countries	Africa
Mobile cellular subscriptions	59	49	32
Fixed telephone lines	19	13	1
Internet users	23	15	4
Fixed broadband subscribers	6	3	0.1
Mobile broadband subscriptions	6	2	0.9

Source: ITU World Telecommunication/ICT Indicators database

Table 8.2 - Estimated Proportion of Households with a TV, by Region, 2009

	Percent
Europe	97
Commonwealth of Independent States	97
The Americas	95
Arab States	82
Asia and the Pacific	75
Africa	28

Source: ITU World Telecommunication/ICT Indicators database

With the steady diffusion of information and communication technologies, African citizens are gradually engaging and consuming media for a variety of private and public uses, not to mention gratifications. Digital media are opening up fresh civic and political prospects and helping to reconfigure the contours of the democratic terrain. The agitations and protests that touched off the revolutions in Tunisia and Egypt during the so-called Arab spring, which helped in dislodging the seemingly impregnable regimes of Ben Ali and Hosni Mubarak, respectively, do testify to the catalyzing effect of digital media. Notable among such digital media are the social networking platforms and their most recognizable brands in Africa – Facebook and Twitter.

For better or worse, the Internet in particular has over a decade matured inexorably into a political force. Whereas this growth cannot be pegged to any single decisive moment, Jenkins and Thorburn (2003) did forecast that early in the last decade "digital democracy will be decentralized, unevenly dispersed, even profoundly contradictory" (2). The predicted contradictions are visible in the way digital democracy has panned out over the years. Authoritarian regimes have, for example, mastered the same technology that has emboldened citizens to ask questions of power and have used it to hit back, literally, at their detractors. Where we used to take virtual space for granted as a safe haven for dissent and critical public discourse, it turns out to have been an unexpectedly short day out for cyber expression.

The unequal pattern of access to the Internet is also manifest in the way the various elites have come to dominate the medium. But while acknowledging the disproportionate control the elite have over the Internet, Powell (2003) contends that information circulating in the Internet-based elite networks typically finds its way to radio, TV, and newspaper audiences the world over: "...so the Internet has an important secondary readership, those who hear or are influenced by online information via its shaping of more widely distributed media, outside of traditional, controlled media lanes of the past" (173). The Internet in Africa deserves more credit as a medium of public discourse than it normally gets. We just have to look no farther than data that attests to the enthusiasm with which the medium is being embraced (see Table 8.3). Regular access to and usage of ICTs (e.g., the Internet and mobile phones) as well as traditional or legacy media (e.g., radio, TV, and newspapers) have been associated with 'cosmopolitanism,' which in turn can stimulate greater political debate and more willingness to accept or tolerate political opposition and dissent (Ismail & Graham, 2009).

Table 8.3 - ICT Growth in Africa and in the World, 2003-2008

	Compound Annual Growth Rate (%)		
	Fixed Telephone Lines	Mobile Cellular Subscriptions	Internet Users
Africa	2.4	47.0	30.6
World	2.5	23.0	17.0

Source: ITU World Telecommunication/ICT Indicators database

Media and democratic citizenship

A well-established view is that media do provide people with the information they need to function effectively as citizens. To be sure, present-day politics is by definition mediated politics. In buttressing this idea, McNair (2000) has gone so far as to assert that:

> Any study of democracy in contemporary conditions is therefore also a study of how the media report and interpret political events and issues; of how they facilitate the efforts of politicians to persuade their electorates of the correctness of policies and programmes; of how they themselves (i.e. editorial staff, management and proprietors) influence the political process and shape public opinion. (1)

Uninformed citizens, goes the argument, cannot hold leaders and governments accountable. Evidently, the structure and organization of modern polities do not allow for direct and regular interaction between citizens and their political representatives. For most citizens, the news media have become the vehicle for political and civic engagement or the lens through which they gather cues to action. On this account, Lewis and Wahl-Jorgensen (2005) argue that the task of journalism is to ferret out public opinion and to speak on its behalf. Yet, ideally, the reporting or media representation of public opinion is only helpful to the democratic process when it motivates citizens to articulate political views of substance as opposed to depicting them as unconcerned with politics and indifferent to civic affairs. However, media critics are disappointed by the reality of what is actually happening:

> All too often, what journalism gives us is thus a series of vague, democratic gestures rather than a dialogue between the people and their political representatives. In this formulation, public opinion is an overwhelmingly reactive, rather than a creative, force. (Lewis & Wahl-Jorgensen, 2005, 106)

The media's failure to raise the quality of public deliberation in general and political discourse in particular has been attributed to a whole bunch of tendencies that have engendered what, to some critics, is effectively a crisis in political communication. These tendencies include dumbing down by trivializing and oversimplifying issues; elevating infotainment at the expense of significant public affairs; privileging elite interests while marginalizing the concerns of those without access to the centers of power; an oversupply of interpretation which drives factual reporting to the periphery; and overplaying the adversarial function of journalism while crowding out the more thoughtful forms of deliberation. Marketisation, commercialization, and commodification are said to be the prime drivers of these practices (McNair, 2000) as journalists and news managers come under unprecedented pressure to profess unwavering commitment to bottom lines and profit margins.

We will pursue the notion of news literacy later on. For starters, we need to elaborate in greater detail the conditions and thinking that have shaped the trajectory of democracy in Africa. Only then shall we be able to clearly situate the role of news literacy in grooming democratically competent citizens especially among the young generation.

Democratization Trends in Africa

The democratic transformation in Africa is often traced to Huntington's (1991) third wave of democratization – a series of trends that either started or took root from 1974 through the 1990s in various countries. The fact that the reforms witnessed in Africa overlapped with and seemingly sprung from larger global developments compelled some observers to view them as microcosms of a worldwide democratic revival "which signaled the historic triumph of liberal democracy" (Szeftel, 1999, 5).

The record on the ground, conversely, demonstrated that the changes were neither as sweeping and clear-cut nor as potent and predictable as theorized. Huntington (1991) himself took note of the patchy pattern left behind by the democratic tide, conceding that as with its forerunners, the third wave was attended by a reverse wave at the macro level; although the reversal never wiped out entirely all the democratic gains of the preceding democratic wave. The political systems and cultures that emerged in post-colonial Africa and those that continued as holdovers from colonialism exemplify the tenuousness of the transitions to democracy. Some of the countries that set out with pluralistic political systems and constitutional orders at independence over the years degenerated into authoritarian states. Typically, what were then dominant-party states (e.g., in Tanzania, Malawi, Kenya, Zambia, Ghana, Sierra Leone) mutated into one-party civilian or military authoritarianism.

In a country such as Uganda, where those excluded from the political process represented a substantial segment of the citizenry, the conflict and instability that ensued had within a decade of independence in 1962 totally swamped civil politics. Ironically, by the onset of the third wave of democratization in the early 1970s, liberal pluralism based on multi-party politics had all but vanished in the former British colonies of Africa except Botswana. These regimes held on to power by largely ditching the values commonly associated with liberal political culture such as civil liberties and the rule of law (Szeftel, 1999).

Leading up to and during the 1990s, pressures to democratize were brought to bear on most African regimes. The mounting demands for political reforms and good governance had international and domestic dimensions, but the results they yielded varied greatly in the magnitude of success and failure within and across different countries. Some regimes allowed only superficial changes as they found alternative strategies to maintain their authoritarian stranglehold on power, while others went down the path of institutional collapse, civil disorder, and violence.

But there were also glimmers of hope in quite a few countries that made notable progress toward democratic rule as in those nations where multi-party elections were successfully concluded, thus ushering a new crop of elites into power. These positive trends made it possible to revisit previous assessments that had painted a predominantly bleak picture of the prospects for democracy in Africa (VonDoepp' & Villalon, 2005).

Rethinking Democracy in Africa

The reality of democracy in the African context therefore requires alternative ways of thinking about the traditional concepts associated with its practice. For example, democratization is conventionally construed in terms of a discrete or bounded process that has precise start and end points. On this notion, the yardstick for democratization is the degree to which a country is relatively democratic, which implies the advancements it has made toward achieving the particular goal of consolidating its democracy. Based on this logic, evaluations of African democracies typically characterize them with labels such as 'feeble,' 'tenuous,' 'quasi,' or 'limited' in order to describe the inadequacies and problematic nature of these democracies (VonDoepp & Villalon, 2005).

Characterizations like these are founded on the assumption that the countries in question are marching to democracy and have only been momentarily interrupted; but that otherwise they are en route to the familiar destination. Assigning countries to specific points on the democratic scale tends to obscure the critical trends that may be unfolding in different nations. VonDoepp and Villalon (2005) concluded from their analysis of the outcomes of democratic experiments in sub-Saharan Africa in the 1990s that most transitional countries had made only limited progress toward establishing democratic norms of fair and open political competition and participation. Furthermore, developments in many of these countries showed that the viability of democracy was doubtful in the long-run.

Yet the gist of what was going on in these countries today defies simple labels and easy categorization. Amidst the setbacks and obstacles to sustaining democracy, many countries have been able to significantly liberalize their politics, thereby dismantling old pillars of authoritarianism and installing new political orders that are also being contested on a recurrent basis. The terms of politics have been transformed in such a way as to create new prospects for political and civic engagement. Citizens have different expectations of their leaders, demanding of them to rely on the performance record of the government, rather than on sectarian appeals, to justify why they should be retained in power. Political discourse itself reverberates with references to the importance and relevance of democracy as a normative principle (VonDoepp & Villalon, 2005).

Against that background, it has been argued that democracy in Africa is essentially in a state of hybridity. This is a condition whereby aspects of democratic behavior and liberal politics co-exist with neo-patrimonial and authoritarian tendencies. Although the persistence of authoritarian rule has to be acknowledged, current developments likewise indicate that there is very scant possibility that the countries now shifting toward more liberalized politics will slide back wholesale into the unbridled dictatorships of an earlier era. "If democracy seems difficult," argue VonDoepp and Villalon (2005), "the liberalization of politics seems irreversible" (6).

Africa so typifies the complex, multi-dimensional, and hybrid nature of democracy that the conventional conceptualization of democratization, which assumes a state of progress toward a specified end, appears to be conceptually insufficient. In many countries, pro- and anti-democratic forces are present simultaneously and the resultant tension calls for a different way of looking at democratization – less in terms of discrete outcomes and more in the context of a complex and variegated system and set of behaviors.

Democracy necessarily implies a certain amount of citizen control over the decisions that impact people's lives (Dahl, 1956). But if this ideal is hard to realize in practice, so much that its tenability is also questionable in countries with long democratic traditions, then its application to Africa's fledgling democracies raises even more problematic issues. As a case in point, the newly democratizing countries had the process certified by holding competitive multi-party elections that in some situations ushered in new leaders, new democratic institutional arrangements, and new configurations of power among political elites. However, the changes these new democracies experienced did not go far enough, if at all, in transferring meaningful power to the people. In many instances, the citizenry remains gravely constrained in its ability to affect policies and to exact serious public accountability from the governing class.

But the myriad and formidable shortcomings of democratization in Africa do not justify discounting the opportunities for enhanced popular control that have been opened up by, and are integral to, the emergent political arrangements. This is because every country that has made a successful democratic transition started out by embracing the kinds of institutional frameworks that, at least in principle, had the potential to produce more accountable forms of government. Here we have in mind institutional frameworks like independent judicial systems, effective parliaments, functional state agencies, as well as independent and pluralistic media.

It would therefore be imprudent to discount the long-term promise and relevance of these fledgling democratic institutions even though they are currently struggling to assert themselves. Without a doubt, they are not yet out of danger from interference by those who see them as inconveniences to their political schemes. If democracy is in part a system whereby elite behavior is subjected to institutional dictates, then the enactment of these new institutions created room for more authentic and wide-ranging democratization.

The larger point here is that the institutions of democracy, however imperfect, can with time and sustained pressure overcome their limitations and grow into confident intermediaries in the process of political reform. Put differently, the mere survival of these systems is a precondition for consolidating democracy. This is notwithstanding the fact that elections have not always fundamentally altered the distribution of power where they have been held; nor have governments always managed to impose serious and durable constraints on office bearers. In many instances, informal and neo-patrimonial political methods take precedence over formal rules in public governance (VonDoepp & Villalon, 2005).

Liberal Applications of Democracy

In general, democratic change in Africa is justified on grounds consistent with core liberal principles that have long been entrenched in Western democracies. However, there are long-standing cautions regarding the transferability of liberal models of democracy to the African context. One of the skeptics has recommended that democratization debates and reforms in Africa should be informed by indigenous political theories (Adam, 1993). Similarly, in looking at the appropriateness of the idea of civil society in Africa given its Western origins, Karlstrom (1999) calls for constant alertness to, and painstaking consideration of, the limitations of such ideas when adapting them to unfamiliar situations.

Clearly primed by such warnings, Berger (2002) submits that much scholarship about the media and democracy in Africa is informed by taken-for-granted liberal pluralist notions of the media as instruments of democracy. On this criticism, the liberal paradigm is inappropriate for Africa because not only is its value being contested in the West, but also because it is based on assumptions that hardly measure up to the political realities and nature of the media environment in Africa.

Hence, liberalism is in some circles considered too narrow to be useful as a universal framework of analysis and as a normative ideal. Yet Adam (1993), despite his own injunction, adopts a concept of democracy that is decidedly liberal to the extent that it revolves around such principles as free competition for power through political parties, political participation via regular and fair elections, as well as civil and political liberties. In the absence of these principles, the integrity of political competition and participation remain doubtful. Adam (1993) endorses this liberal definition of democracy because it "captures many of the demands made by the current African opposition groups, parties, and movements" (500).

Cognizant of the criticisms leveled against Western democratic models, Sandbrook (1988) nevertheless contends that Africa's objective conditions do not necessarily preclude the feasibility of liberal democracy because "the structural-determinist thesis seems indefensibly negative" (251). He points to some possible outcomes of liberal democracy, such as the potential to provide defenses against tyranny and despotism and to open up political space to enable those opposed to the dominant groups to assert their claims.

To be sure, many of the positive initiatives and advances toward democratization on the continent have been secured by and large through the creation, revival, and promotion of those institutions, principles, practices, and behaviors typically associated with the liberal model of democracy. As they went through their democratic transitions, many countries across the continent witnessed vibrant agitations for political goods whose demand was inspired by essentially liberal values, for example: elections held regularly and in predictable cycles; competing political groups; independent parliaments and judiciaries; autonomous civil societies and public spheres; accountability and transparency in government; protection of human rights; freedom of expression; free media; the rule of law; equal protection under the law; the right of access to public information; and checks and balances on power (see also Diamond, 1997).

The disconnect between theory and practice notwithstanding, experiences with democracy in Africa demand that we take seriously the appeal, if not of all but certainly of some aspects, of the liberal model however uneasy their accommodation might be (Makinda, 1996; Decalo, 1992; Sklar, 1983; Kawonise, 1992; Szeftel, 1999; Wiseman, 1990).

And there are compelling reasons to heed that demand. Results from research done between 1999 and 2000 – at the finale of a decade marked by historic economic liberalization and political reforms – confirmed that popular conceptions of democracy in Africa were indeed liberal (Bratton & Mattes, 2000). According to data from the first round of Afrobarometer opinion surveys, 34 percent of the 10,398 respondents interviewed in Botswana, Ghana, Malawi, Namibia, Nigeria, and Zimbabwe associated democracy with civil rights and personal freedoms more frequently than with other conceptions of democracy. This figure represented an understanding of democracy that emphasized individual rights as opposed to the less than 0.1 percent of respondents who referred to group rights. This led to the conclusion that:

> Contrary to those who would have us believe that Africans conceive of democracy and associated rights in a different way than Westerners, our survey respondents are telling us that they place individual rights uppermost. And, to the extent that they claim such rights as a means of resisting repression at the hands of an authoritarian ruler,

Africans are beginning to think more like citizens of a constitutional state than clients of a personal patron (Bratton & Mattes, 2000, 4-5).

News Literacy and Democratic Citizenship

The Afrobarometer cross-national research program, also the most compre-hensive work yet on public opinion in Africa, has turned up telling results about the content of and influences on mass attitudes toward democracy. Spe-cifically, the research has demonstrated that "cognitive awareness" of public affairs and "performance evaluations" of regimes were singularly critical to public opinion formation. Following what is defined as a "learning approach" founded on the assumption of "knowledge and experience as the key determi-nants of public opinion," the research has empirically clarified Africans' con-ceptions of democracy; the criteria by which they evaluate regime performance; the nature of their support for regimes; their expectations of democratic leadership; and how they learn about politics (Bratton, Mattes, & Gyimah-Boadi, 2005, 44).

Given the evidence at our disposal, we now have a better grasp of public opinion on issues of collective interest to African citizens. Most importantly, we have a basis for interrogating the nature of, as well as the factors that influ-ence, democratic citizenship in Africa. Obviously, democracy needs democrats to blossom – people with democratic convictions or mind sets that favor de-mocratic norms. Gyimah-Boadi and Attoh (2009, 1) have used data from pub-lic attitude surveys "to explore the extent to which Africans are orienting their attitudes and behaviour in the manner expected of citizens in a democratic society." The Afrobarometer carried out the surveys in 19 countries in 2008. The notion of "democratic citizenship" as applied in their analysis of the Afri-can case has three elements, that is: whether people in Africa "hold democ-ratic attitudes and values," "demonstrate political knowledge and engagement," and "exhibit democratic behaviours via civic participation."

The conclusion drawn from the assessment of the trends was one of a "mixed picture." Whereas there were promising signs in all dimensions of de-mocratic citizenship, there were at the same time worrying indicators across the board regarding "the depth and extent of democraticness among African citizens" in the 11 countries where trends from 1999 to 2008 were compared. The verdict was that democratic citizenship in Africa remained "relatively weak." On the plus side, however, the direction of the trends was heartening. "The positive trends indicate that democratic citizens are beginning to emerge

in Africa to complement the democratic structures and processes that are being established in the countries surveyed" (Gyimah-Boadi & Attoh, 2009, 2).
Evidently, there is still much work to do to entrench the attributes of democratic citizenship among Africans. And it is in the service of this cause that
news literacy acquires its relevance and gains its vitality as a driver of deeper
democracy.

For our purposes, the term 'news literacy' is a very specific rendition of its
mother concept, 'media literacy' (for definitions see Mihailidis, 2009, and
others elsewhere in this volume). Media literacy is generally associated with the
acquisition of a range of skills including critical thinking, problem solving,
personal autonomy, as well as social and communicative skills (Martinsson,
2009, 3). These skills are considered essential as motivators of informed and
active citizenship.

Journalism at its core is about what Adam and Clark (2006) refer to as
"news judgment." They describe news judgment as "a form of vision, a way of
knowing the here and now, that leads reporters and editors to notice events
and things that are likely to matter in a democratic society" (46). When applied to public life and civic engagement, *news literacy* is to be understood as a
way of knowing about the social, cultural, economic, and political forces operating in the citizen's local, national, and global environments.

The emergence of the practice of 'citizen journalism' (Moeller, 2009;
Banda, 2010), regardless of the many contending interpretations of what the
concept itself means and its association with crisis reporting (Allan, 2010),
underlines the importance of citizens developing their news judgment capabilities. These capabilities are both intellectual (being able to discover and analyze information) and practical (being able to manipulate media and to create
and disseminate content).

Our exploration of the links between democratization and the media
demonstrates that efforts to deepen democracy in Africa have a lot to gain by
being grounded in news literacy. As we have pointed out, the attributes of
democratic citizenship imply a citizenry that is routinely informed about public affairs, is active in civic life, and is engaged with the political process. Many
surveys have shown that in Africa too, the media especially radio are the primary sources of information about public and civic affairs. The main question
of interest going forward is not whether Africans have access to media or what
they use them for. The primary issue has to do with the cognitive capital they

bring to their consumption of media and the intellectual skills they employ to digest and distill the information they get from the media.

The role of the media in democratization in Africa is largely under-appreciated despite the rhetoric of governments. The governing elite value the media only to the extent that they do their bidding. The relentless attacks on media freedom point to the fragile existence of the media as a democratic institution in its own right. These attacks come in many forms: physical harm; intimidation by public officials, politicians, and security agencies; illegal detentions; unconstitutional charges and prosecutions; the seemingly benign but no less insidious attempts to co-opt journalists into serving as state agents; threatening reporters, news sources, and commentators with labels such as state saboteur; and the constant attempts to narrow media space and freedom of expression through opportunistic legislation.

Given that the media in their various forms are woven to varying degrees into citizens' daily encounters with public life, there is growing recognition globally of the necessity to integrate news literacy into the processes through which people learn to become democratic citizens. For instance, the UN-Alliance of Civilizations (2009) has tracked the work going on worldwide toward this goal. It argues that media literacy is critical to the acquisition of critical thinking skills both among the youth and in society generally. Furthermore, media education conveys and nurtures the vital knowledge and analytical tools that would enable audiences to mature into "autonomous and rational citizens" who are capable of engaging the media as informed consumers. This capability is particularly crucial with regard to news. In June 2008, UNESCO convened an international meeting of experts in Paris who came up with a strategy for teacher training a curriculum on media and information literacy.[4] UNESCO's other initiatives in this area include a media education kit (UNESCO, 2006).

News literacy among citizens will provide an institutionalized avenue to counter the many and frequent affronts on the media described earlier. Citizens who hold strong convictions about the role of the media in consolidating democracy and in making governments more responsive to the interests of the citizenry will be less tolerant of attempts to undermine their work and existence. Moreover, news literacy will arm citizens with the information and arguments they need to voice their convictions. Most critically, it will motivate them and give them the confidence to initiate action or to seek out opportunities to participate in collective action for the common good.

Thankfully, there is growing body of work on news literacy and all its variants such as media literacy and media education. Gillmor's (2010) book aptly titled *Mediactive* is essentially a citizen's do-it-yourself manual on news literacy with tips for both consumers and producers of media and information.

Conclusion

This chapter has demonstrated the connection between democracy and the media as experienced in the African context. Using that connection as its premise, the discussion has underlined the need to consider news literacy as part of the foundation of democracy in Africa as it should be elsewhere. The news media encapsulate the whole gamut of public and civic life, and no account of the political and democratic processes is complete without factoring the media into the equation. Active engagement with the news media is the primary means by which citizens will gain the knowledge that will give them a significant voice in public affairs and that will also inspire them to get involved. The discussion has also drawn attention to existing templates that should offer a useful starting point for integrating news literacy in education systems and civic education programs throughout the continent. Before this happens, the relevant arguments and the necessary connections between news literacy and civic outcomes have to be made. That, hopefully, has been the contribution of this chapter.

REFERENCES

Adam, G.S. & Clark, R.P. (2006). *Journalism: The Democratic Craft*. Oxford: Oxford University Press.

Adam, H.M. (1993). Frantz Fanon as a Democratic Theorist. *African Affairs* 92, 499-518.

Afrobarometer (2010). Kenyans and the Coalition Government: Disappointment in spite of Relative Peace. Afrobarometer Briefing Paper No. 91, August. Retrieved on 16 June 2011 from http://www.afrobarometer.org/index.php?option=com_docman&Itemid=37

Allan, S. (2010). *News Culture*, 3rd ed. Maidenhead, UK: Open University Press.

AudienceScapes (2010). Radio, Mobile Phones Stand Out in Africa's Media/Communication landscape. Retrieved on 16 June 2011 from:
http://www.audiencescapes.org/sites/default/files/AScapes%20Briefs%20New%20Media_Final.pdf

Banda, F. (2010). *Citizen Journalism and Democracy in Africa*. Grahamstown, Souht Africa: Highway Africa.

Berger, G. (2002). Theorizing the Media-Democracy Relationship in Southern Africa. *Gazette* 64(1), 21-45.

Bratton, M. & Mattes, R. (2000). Democratic and Market Reforms in Africa: What 'the People' Say. *Afrobarometer* Paper, No. 5. Retrieved on 16 June 2011 from:
http://www.afrobarometer.org/index.php?option=com_docman&Itemid=39&limitstart=100

Bratton, M., Mattes, R.B., & Gyimah-Boadi, E. (2005). *Public Opinion, Democracy, and Market Reform in Africa*. Cambridge, UK: Cambridge University Press.

Dahlgren, P. (1995). *Television and the Public Sphere: Citizenship, Democracy and the Media*. London: Sage Publications.

Decalo, S. (1992). The Process, Prospects and Constraints of Democratization in Africa. *African Affairs* 91(362), 7-35.

Diamond, L. (1997). *Prospects for Democratic Development in Africa*. Palo Alto, CA: Stanford University Press.

Gillmor, D. (2010). *Mediactive*. Lulu.com.

Gyimah-Boadi, E. & Attoh, D.A. (2009). Are Democratic Citizens Emerging in Africa? Evidence from the Afrobarometer. *Afrobarometer Briefing Paper* No.70. Retrieved 16 June 2011 from: http://www.afrobarometer.org/index.php?option=com_docman&Itemid=39&limitstart=100

International Telecommunication Union (2009). *Information Society Statistical Profiles 2009: Africa*. Geneva, Switzerland: ITU.

Ismail, Z. & Graham, P. (2009). Citizens of the World? Africans, Media and Telecommunications. *Afrobarometer Briefing Paper* No. 69. Retrieved 16 June 2011 from: http://www.afrobarometer.org/index.php?option=com_docman&Itemid=39&limitstart=1 00

Jenkins, H. & Thorburn, D. (2003). Introduction: The Digital Revolution, the Informed Citizen, and the Culture of Democracy. In H. Jenkins & D. Thorburn (Eds.), *Democracy and New Media*, 1-17. Cambridge, MA: MIT Press.

Karlstrom, M. (1999). Civil Society and Its Presuppositions: Lessons from Uganda. In John L. Comaroff & Jean Comaroff (Eds.), *Civil Society and the Political Imagination in Africa*,104-123. Chicago: The University of Chicago Press.

Kawonise, S. (1992). Normative Impediments to the Democratic Transition in Africa. In B. Caron, A. Gboyega, & E. Osaghae (Eds.), *Proceedings of the Symposium on Democratic Transition in Africa*, 129-140. Ibadan, Nigeria: Centre for Research, Documentation and University Exchange (CREDU).

Lewis, J. & Wahl-Jorgensen, K. (2005). Active Citizen or Couch Potato? Journalism and Public Opinion. In S. Allan (Ed.), *Journalism: Critical Issues*, 98-108. Maidenhead, UK: Open University Press.

Makinda, S. (1996). Democracy and Multi-Party Politics in Africa. *The Journal of Modern African Studies* 34(4), 555-573.

Martinsson, J. (2009). *The Role of Media Literacy in the Governance Reform Agenda*. Washington, DC: The World Bank.

McNair, B. (2000). *Journalism and Democracy: An Evaluation of the Political Public Sphere*. Abingdon, UK: Routledge.

Meyer, P. (2003). The Proper Role of the News Media in a Democratic Society: Is It Enough to Simply Cover the News. In J. Harper & T. Yantek (Eds.), *Media, Profit, and Politics: Competing priorities in an open society*, 11-17. Kent, OH: The Kent State University Press.

Mihailidis, P. (2009). *Media Literacy: Empowering Youth Worldwide*. Washington, DC: Center for International Media Assistance.

Moeller, S. D. (2009). *Media Literacy: Citizen Journalists*. Washington, DC: Center for International Media Assistance.

Powell III, A. C. (2003). Democracy and New Media in Developing Nations: Opportunities and Challenges. In H. Jenkins & D. Thorburn (Eds.), *Democracy and New Media*, 171-177. Cambridge, MA: MIT Press.

Sandbrook, R. (1988). Liberal Democracy in Africa: A Socialist-Revisionist Perspective. *Canadian Journal of African Studies* 22(2), 240-267.

Sklar, R. L. (1983). Democracy in Africa. *African Studies Review* 26(3), 11-24.

Szeftel, M. (1999). Political Crisis and Democratic Renewal in Africa. In J. Daniel, R. Southall, & M. Szeftel (Eds.), *Voting for democracy: Watershed Elections in Contemporary Anglophone Africa*, 1-18. Aldershot, UK: Ashgate.

UN-Alliance of Civilizations (2009). *Mapping Media Education Policies in the World: Visions, Programmes and Challenges*. New York: UN-Alliance of Civilizations.

UNESCO (2006). *Media Education: A Kit For Teachers, Students, Parents and Professionals*. Paris: UNESCO. Retrieved 17 June 2011 from:
http://unesdoc.unesco.org/images/0014/001492/149278E.pdf

VonDoepp, P., & Villalon, L. A. (2005). Elites, Institutions, and the Varied Trajectories of Africa's Third Wave Democracies. In L. A. Villalon & P. VonDoepp (Eds.), *The Fate of Africa's Democratic Experiments: Elites and Institutions*, 1-26. Bloomington, IN: Indiana University Press.

Wiseman, J. A. (1990). *Democracy in Black Africa: Survival and Revival*. New York: Paragon House Publishers.

Conclusion -
News Literacy and The Courage to Speak Out

SUSAN MOELLER

University of Maryland, College Park, USA

Introduction

One morning, just before I sat down to finish writing this conclusion on the critical importance of news literacy across the world, I listened to an interview by National Public Radio reporter Steve Inskeep with Asma Jehangir, a human rights advocate and lawyer in Pakistan. Inskeep (2011) reminded his American listeners that Pakistan is a country where "vastly more is spent on defense than, say, education."

Inskeep began the interview by asking Jehangir about a recent TV appearance of hers that "received quite a bit of attention. "That's not the first time I've spoken up," replied Jehangir, "but this was a live program so there could be no censorship on it. And what I said is that the military leadership is — I used the word 'duffer.' And I'm sorry I used the word duffer — I should have said dangerous duffers."

Inskeep followed up: "Is there a limit to how much you can criticize the military?" Jehangir responded, "They don't like criticism — well, too bad for them. I don't like to be killed in violence." "Is it dangerous to speak like that?" asked Inskeep. "I'm sure it is," she replied. "But we all live dangerous lives here."

An hour or two later, I spoke with Naziha Ali, a Karachi-based journalist who noted that "Asma Jehangir is fearless. In fact, it wouldn't be an exaggeration to say that she's the voice of conscience in Pakistan. The excerpt from the TV program in which she condemned the military as a ruthless occupation force went viral within a few hours of the program's airing."

Then, that same afternoon, news broke that Syed Saleem Shahzad, a Pakistani reporter working for the Hong Kong–based *Asia Times Online* and *Adnkronos International*, had been murdered. He had disappeared on his way to a television interview in Islamabad two days before where he had been scheduled to talk about his recently published story about al-Qaeda's infiltration of Pakistan's navy.

Shahzad's wife had called Human Rights Watch after he had failed to show up at the TV interview or return home. HRW then raised an alarm over his disappearance, citing a "reliable interlocutor" who said that the country's powerful Inter-Services Intelligence spy agency [ISI] had abducted him.

Nazhia Ali confirmed that the reports of his torture and death were immediately "all over the news, FB [Facebook], Twitter etc. Horrific thing to happen and people are very upset. Although Saleem Shahzad's abduction, torture and murder is typical of the agencies' modus operandi, their brazen indifference to public opinion has been highlighted as never before. This was a journalist who had just reported on an extremely important story, not some poor Baloch nationalist who could safely be eliminated without much fuss in the media."

"The Pakistani media," Ali noted, "is not as free as one might assume. There are certain no-go areas where reporting is concerned (e.g. policy matters vis-à-vis India, the nuclear issue, the Balochistan insurgency and the nexus between the ISI and Islamic militants) and journalists like Saleem Shahzad who refuse to be muzzled pay the ultimate price."[1]

News Literacy Is Integral to Press Freedom

According to the Committee to Protect Journalists, Pakistan is the deadliest country in the world for journalists. And Reporters Without Borders ranks Pakistan 151 out of 178 countries for press freedom.

It's hard to imagine a country that more needs its population to be news literate, that has a greater need for engaged, informed and active citizens. Consider the problems: endemic poverty; a crisis of education, especially for girls; repeated disasters, such as the 2005 earthquake that killed 75,000 and the 2010 floods that inundated one-fifth of Pakistan's total land area; internal violence, including suicide bombings and assassinations; cross-border conflict with India; and international forces fighting against the Taliban and al Qaeda forces in Afghanistan, Khyber Pakhtunkhwa and FATA (Federally Administered Tribal Areas).

In a country where almost 1 in 10 children does not live to age five – with most dying from diarrhea, pneumonia or vaccine-preventable diseases – the parents needs to be news literate about basic health issues. In a country where in the recent past the president has suspended the Chief Justice of the Supreme Court and the Constitution, the public needs to be news literate about the rule of law. In a country where major domestic and international media outlets, such as GEO TV, the BBC and CNN, have at times been shut down, citizens need to be news literate — not just to defend the independence of the press, but to know how to access and use alternative sources of news, as well as how transmit their own information.

Over 100 million Pakistanis own mobile phones, according to the Pakistan Telecommunication Authority[2] – and that's the platform by which most Pakistanis at all economic levels get and exchange any news that might be considered "no-go." As Pakistani blogger Zaira Rahman Sheikh (2011) notes, "It is not an exaggeration but even the maids, drivers, guards and beggars on the streets own a mobile set or two.... So, it is no longer surprising to see [F]acebook status updates, instant picture uploads and tweets being followed by the second via cell phones because things have generally become much easier today.... What was earlier considered a luxury is becoming a necessity of life."

Mobile phones, in other words, are more than a social tool to link individuals and they are more than the industrial growth sector of digital access. For the over half of the population of the world who live in countries that do not have a free press, mobile phones have become essential tools for survival. (Freedom House, 2010). Mobile phones give those citizens their closest approximation to free and independent media. Information received through text messages, gives those publics the power to hear unfettered news.

Mobile phones do not by themselves give people their "First Amendment" rights – or their rights under Article 19 of the Universal Declaration of Human Rights – but mobile phones go some way towards assisting individuals to assert their rights to free speech and a free press, as well as their right to assemble. If social networks are today's public square, the mobile phones in everyone's hands enable everyone to be a town crier.

Jan Chipchase, a scientist and former design strategist for Nokia who now travels the world for *frog*, confirms how mobile phones have become necessities of life: "If you wanted to take phones away from anybody in this world who has them, they'd probably say: 'You're going to have to fight me for it. Are you going to take my sewer and water away too?' And maybe you can't put communication on the same level as running water, but some people would. And I think in some contexts, it's quite viable as a fundamental right."

Seen in this light, it's not a coincidence, but rather a sign of the times that Asma Jehangir's outspoken interview went viral through individuals passing it on via texts and email, and that the breaking news and commentary about

Saleem Shahzad's abduction, torture and murder were passed on through Twitter and Facebook posts. Noted activist Bill Shore (2009): "Social media can't ensure social justice. But it can affect the invisibility that is the first barrier to achieving it."

Press Freedom Is Integral to Civil Society

The example above is but one instance where the need for news literacy is fundamental to the foundations, frameworks, and structures of civil societies around the world. This book has combined scholarship, theory, pedagogy and practice to make the case for news literacy in the newsroom and the classroom, aimed at empowering civic society and embracing democracy. The chapters above support a fundamental need to better prepare tomorrow's reporters and citizens, for lives of active civic inquiry and tolerance for diverse viewpoints. For as media and information become even more central to the functioning of communities large and small, teaching citizens how they can responsibly and meaningfully contribute to society will ultimate improve civil society and its institutions.

The authors in this book make a uniform case that news literacy strengthens independent media. The authors argue that news literacy teaches audiences why it is in their interest to support a diversity of news outlets and hear from a plethora of individual voices. Mobile technologies, participatory social media, both joined with professional and citizen journalism, are remaking what the global public knows and thinks about the world. In Chapter One, Stuart Allan alludes to this need in the context of civic protest and resistance, while in Chapter Seven Jad Melki shows how such understandings of mobile technologies and news flow can be enhanced through pedagogies that embrace media literate outcomes. Further, in Chapter 8 George Lugalambi stresses individual and diverse voices as the cornerstone for civic democracy throughout Africa.

That is why there is such a need to teach – and continually update the teaching of – news literacy around the world. Students and adults need to learn how to access information, use media across platforms, and responsibly contribute news and information to local, national and international audiences. Whether Cairo or Chicago, Uganda or the United Kingdom, the core competencies of news literacy, as conceived in the chapters of this book, are critical to address immediate as well as endemic problems – not the least of which is strengthening journalism across the globe.

The authors in this book make the dramatic claim that news literacy is key to solving the problems of the 21st century world. They are not the first to identify the power of information: "When key constituencies have information, media can have a positive effect on corruption, political turnover and media capture," observed London School of Economics professor Tim Besley in 2002. But the authors here, gathered from all over the world, make the further claim that policy makers, not just educators, need to see news literacy as an essential competency for their citizens.

News literacy can encourage citizens to become informed and to become activists for civil society and improvements in their quality of life. Nobel Prize-winning economist Amartya Sen gave an example of that phenomenon in his 1982 Coromandel Lecture: "India has not had a famine since independence, and given the nature of Indian politics and society, it is not likely that India can have a famine even in years of great food problems. The government cannot afford to fail to take prompt action when large-scale starvation threatens. Newspapers play an important part in this in making the facts known and forcing the challenge to be faced."

But before newspapers – or today other kinds of media, too – can serve those functions, they have to be seen as truth-tellers. And when media are handmaidens of the state or of some other entrenched elite interest, the information being disseminated is suspect. Oxford economist Paul Collier, for example, noted that a key barrier to building a more civil society not only in nations at risk, but in existing democracies, as well, is that "Governments have realized that they can evade accountability while still having elections as long as they muzzle the press or buy the press."

This book has confirmed what Collier has written: that citizens cannot elect their best representatives, accurately monitor industry or best judge how to foster healthy development because they "encounter a double layer of difficulty. They are starved of information and they don't even know how much trust to place in information sources that are available to them." We have learned, Collier has said, that "elections only work if we complement them with an informed society."[3]

As both Manuel Guererro and Raquel San Martín note, too many environments, across the globe, are rife with censorship, intimidation and monopoly ownership, all factors that depress fair and accurate coverage of issues and events. In such situations, Guerrerro and San Martín agree with Collier, elections do not perfunctorily bring good governance, corporate accountability and strong economic development because citizens are uninformed.

News literacy education makes citizens mindful of what information is available and who is served by that news – once educated, citizens become aware of the quality and value of the news. When members of the public are news literate they understand how essential access to independent news and information is to the exercise of citizenship and how essential media are to bringing transparency and accountability to government and the corporate sector. The question then becomes, how can we best prepare citizens to become mindful, skeptical and aware of the value of news for their rights and responsibilities to society? Author Stephen Reese in his chapter on pedagogical developments and Constanza Mujica in her work on how case studies can encourage cross-cultural understanding both wrote about the need for news literacy to be integrated not just into educational systems, but into communities and homes, as well.

Civil Society Is Empowered by New Media

Beyond the theoretical connections among news, democracy and citizenship explored in Part One, this book investigated the new social media platforms and mobile technologies changing how individuals can relate to news and elites. Citizens around the world have always had a need to understand the media they receive – to mention just one historical example, consider the manipulation of national and international audiences by government propaganda during the First and Second World Wars. But today the rate of technological change means that citizens need the skills to evaluate government propaganda and media spin on constantly changing platforms. In Chapters One, Two and Six, Stuart Allan, Manuel Guerrero, Moses Shumow and Sanjeev Chatterjee, all addressed how the new digital technologies are empowering citizens:

- Today's new tools make it possible to WITNESS the world in different ways. More people can cover events and issues than ever before – the dramatic spread of mobile technologies, and the lower cost of taking, producing and disseminating news mean that everyday citizens as well as journalism professionals are now core contributors to the web of information out there. But the permutations about who is a "journalist" now go beyond the binary choice of citizen v. professional.

Curation tools, such as Storify, allow anyone to search through reams of information to create a useful online "article" from the bits of information collected from the proverbial "man on the street." [4] And crowd-sourcing tools, such as Storiful and Yogile, allow the contributions of many people to be gathered quickly together into a coherent narrative or visual whole.

- Today's new tools make it possible to REPORT ON the world in different ways. Professional journalists report across platforms, literally 24/7. Their stories can appear in print and on radio and TV, as well as online, and via a handheld. And news on those separate platforms reaches distinctive and more diverse audiences. What can or should be told differs by platform, too. A story, even about the same event and told by the same reporter, will be appreciably different if it is read in a newspaper (whether in newsprint or on an e-reader), listened to on the radio (while driving in a car or from a downloaded MP3 file), watched as a video (on broadcast TV or through a streaming online site), or inhaled in short chunks (through an email listserv, as Facebook posts, 140-character tweets or scrolling IMs).

- Today's new tools make it possible to SEE the world in different ways. Still and video cameras are more compact than they ever were, less expensive than they ever have been, and in use by more people than ever before. Images and video can be almost instantly uploaded to professional and open-access sites, so that the world can see what the camera operator sees. And today's software – from Photoshop to Instagrams – that accompanies the hardware means that images, even in camera, can be easily and immediately tailored (or manipulated) to appeal to a particular audience or to deliver a particular message.

- Today's new tools make it possible to VISUALIZE global patterns, changes and trends. Visualization software allows journalists to find and research relationships in data as never before. Data that would have taken teams of researchers days or weeks to parse can now be mined and shaped in seconds. Data visualization tools allow their users to see problems and opportunities where just a mass of words and numbers existed. Such tools allow for greater insight into and oversight of both political and economic issues as well as provide opportunities for greater transparency about the assessment and analysis of data.

- Today's new tools make new business models possible for PROFESSIONAL as well as CITIZEN JOURNALISTS. E-commerce, micro-payment and crowd-sourcing technologies have all given rise to new possibilities for underwriting journalism. As yet there appears to be no "silver bullet" that shoots to the heart of the how-do-we-pay-for-quality-journalism conundrum. But new as well as hybrid modules suggest that there can be innovation in the business of journalism, as well as the reporting of it.

New Media Are Expanding How We Get News – And What News We Get

Today, no matter where we live in the world, we are swimming in a sea of media. But like the oxygen that we breathe, we are often oblivious. Music flows from cars and busses waiting at stoplights and through the walls of neighbors' apartments. Televisions blare from walls in restaurants and bars, gyms, public transportation facilities and other places of gathering. Newspapers and tabloid tear sheets announce their headlines from bins and kiosks on the street. Computers buzz in workplaces and classrooms. Mobile phones ring and ping in our pockets and handbags.

And most of the young today wouldn't have it any other way. Paul Mihailidis, Editor of this book, notes in in his introduction the sheer ubiquity and centrality of mobile technologies in the lives of youth today. Said one British college student in response to a 2011 Salzburg-ICMPA global study on media use: "We feel the need to be plugged in to media all day long. Our lives basically revolve around it. It is the way we are informed about news, about gossip, the way we communicate with friends and plan our days" (Moeller, 2010).

The sheer ubiquity of ambient media has affected how we get information. Until recently, the rule was that the curious searched for news. But now the news finds its audience. Unlike their pre-Web 2.0 predecessors, people today – perhaps especially the young – are less and less frequently going in search of news. Young adults now squat in place on Facebook pages, Twitter accounts, chat platforms and e-mail, gathering their news as it appears on their own pages.

A decade or more ago there was much public hand-wringing about a then-newly observed phenomenon: the Internet was paradoxically limiting users' intake of information. Despite the exponentially increasing amount of news and information accessible online, librarians, professors, journalists and parents worried that the marvelous opportunities for serendipitous discovery of new information when browsing a library's shelves or paging through a newspaper were being lost.

Internet users weren't stumbling over provocative books or articles that expanded or challenged their understanding of the world because, so research suggested, users went to pages and sites that told them exactly what they wanted to hear. Using bookmarks and other electronic means to demarcate where they wanted to go, users commonly visited specific sites they had pre-identified for news and entertainment.

Essentially, so the argument went, virtual ruts were emerging analogous to the "worn highways" that Henry David Thoreau warned against in his conclusion to *Walden*:

> "I had not lived there a week before my feet wore a path from my door to the pond-side; and though it is five or six years since I trod it, it is still quite distinct. It is true, I fear that others may have fallen into it, and so helped to keep it open. The surface of the earth is soft and impressible by the feet of men; and so with the paths which the mind travels. How worn and dusty, then, must be the highways of the world, how deep the ruts of tradition and conformity!"

But today, according to the 2011 Salzburg-ICMPA global study that asked over 1000 students on five continents how they use media, young users are no longer travelling the same virtual ruts as before. Students rarely go prospecting for news at mainstream or legacy news sites, the study found. They inhale, almost unconsciously, the news served up on the sidebar of their e-mail account, posted on friends' Facebook walls or delivered by Twitter.

Librarians, professors, journalists and parents may still bemoan this generation's loss of initiative and the kind of active curiosity necessary to gather information in an unwired world, but students today are plugged into news via their friends in unprecedented ways. No matter where they live, students observe that they are inundated with information coming via mobile phones or the Internet – text messages, social media, chat, e-mail, Skype IM, QQ, Weibo, RenRen and more.[5]

The consequence of the non-stop deluge of information, according to the Salzburg-ICMPA study's college students, means that students have neither the time nor the inclination to follow up on even major news stories. Most students reported that a short text message from a friend is sufficiently informative for all but the most personally compelling events. Said one American:

"We are so used to easily accessing anything that we need at anytime that we want. Journalists' roles are diminishing because the new generation is not interested in newspapers or magazines. All we want to be able to do is click a mouse and be exposed to any information. If it is not interesting enough, we will just move onto a different site with another click."[6]

For daily news, students are now headline readers via their social networks. It's only for major news stories that students' reported that they felt moved to go beyond Facebook and other social networking sites. Noted an American: "I got the news of Bin Laden's death on Facebook. I had heard it from a couple of people on my Facebook chat. A lot of people find their news out this way, and if it's something that is a big deal, later they turn on their TV to hear more about the details." A second student agreed: "I also heard about Osama Bin Laden's death through Facebook. I was driving the night it occurred and I had my iPod playing through my car radio. Out of habit I randomly checked my Facebook on my phone and saw various status's in my news feed about his death. I then turned my radio on to confirm this and also to get more detailed information."

Students, in other words, aren't desperate to read or surf to *The New York Times*, the BBC or their equivalents. And because Facebook, Twitter, Gmail and their counterparts are increasingly the way that students receive their news and information, many youth today are cavalier about the need for traditional news outlets: "We are used to having information about everything on the planet and this information we have to have in an unbelievable time," said a student from Slovakia. "Our generation doesn't need certified and acknowledged information. More important is quantity, not quality of news."

It wasn't that students in the Salzburg-ICMPA study reported a lack of interest in news – they just reported that they cared as much about what their friends were up to as they cared about local and global news. "I felt a little out of touch with the world," reported a student from the UK after going without media for 24 hours, "and craved to know what was going on not only in worldwide news, but with my friends' everyday thoughts and experiences, posted in statuses, tweets and blog posts daily."

Conclusion: How We Get News – And What News We Get – Makes It Imperative That We Are News Literate

The essence of *News Literacy: Global Perspectives for the Newsroom and the Classroom* is that it situates the need for understanding information at the center of civic and democratic identity in the 21st century.

Each contributor to the book harnesses their experience and expertise to help create a journey in news and media literacy that aims to connect the fundamentals of news production, dissemination and reception with technology, pedagogy and culture. The fact that today's youths are sitting like spiders in the middle of a web, often content with consuming what flies by, carries serious social and political consequences in an era where Facebook and Twitter have become the media of choice for governments and politicians for public outreach—can the messages of the spin doctors be distinguished from the information passed on by regular citizens and the reporting of professional journalists? The passivity of a generation also carries consequences when independent news outlets are routinely censored and attacked and when journalists such as Syed Saleem Shahzad are threatened and killed. As the authors in this book detail, news literacy is needed now more than ever.

News literacy needs to be taught as part of the core curriculum in universities around the world. News literacy not only teaches the skills of critical thinking and analysis, it teaches the value of news and information, the power of media messages, the role that the public can – and should play – in setting the public agenda. News literacy is nonpartisan. News literacy programs do not direct their audiences how to be engaged and certainly do not tell them how to vote. But news literacy does prepare the public for active and inclusive roles in information societies.

Without news literacy education there is too little pressure and a too minimal audience for quality journalism on any platform. News literacy teaches citizens about the role of media in their lives – how to distinguish between fact and fiction, credible and non-credible sources, important and unimportant information, and how to mindfully navigate multiple platforms for personal and professional purposes without becoming toxically overwhelmed and distracted.

News literacy—like the traditional kind of ABCs literacy—is also about access to information. News literacy teaches the global public how to find the news they need, how to evaluate what they see, hear and read, how to make sense of the problems that are identified and the solutions that are on offer, and how to engage with others proportionately and responsibly. News literacy teaches the public to recognize the value of quality news – and thereby increases the demand for it.

The goal of news literacy is to give people the power to use their rights of free expression, to defend their access to information, to secure their participation in the process of governing, to help all voices be heard. In a myriad of ways, with multiple voices, this book argues that news literacy does not just open books. It opens doors, opens minds, opens worlds.

REFERENCES

Besley, T. & Burgess, R. (2002). "The Political Economy of Government Responsiveness: Theory and Evidence from India." *The Quarterly Journal of Economics*, 117(4), 1415-1451.

Freedom House. (2010). "Freedom in the World 2010: Erosion of Freedom Intensifies." *The Freedom House*. Washington DC: Retrieved on 4 June 2011 from http://www.freedomhouse.org/template.cfm?page=505

Inskeep, S. (2011). "Criticizing Pakistan's Military: Dangerous, as Is Life." *National Public Radio*, Washington, DC, 13 May. Retrieved on 4 June 2011 from http://www.npr.org/2011/05/31/136800669/criticizing-pakistan-military-dangerous-as-is-life

Moeller, S. (2010). "24 Hours without Media." *International Center for Media and the Public Agenda, University of Maryland*. Retrieved on 22 April 2011 from http://theworldunplugged.wordpress.com

Sheikh, Z. R. (2011). "The Telcom Revolution! Pakistani Media: The Way Things Are 1-6." *Pakistan Media: The Way Things Are*, Retrieved 4 June 2011 from http://pakistanimedia-thewaythingsare.blogspot.com/2011/03/telcom-revolution-pakistani-mediathe.html

Shore, B. (2009). "When Social Media and Social Justice Intersect." *The Huffington Post*, 13 July. Retrieved on 4 June 2011 from http://www.huffingtonpost.com/billy-shore/when-social-media-and-soc_b_231018.html

LIST OF CONTRIBUTORS

STUART ALLAN

Stuart Allan is Professor of Journalism in the Media School, Bournemouth University, UK. His authored books include *Digital War Reporting* (co-authored with D. Matheson, 2009), *Nanotechnology, Risk and Communication* (co-authored with A. Anderson, A. Petersen and C. Wilkinson, 2009), *News Culture* (third edition, 2010) and *Keywords in News and Journalism Studies* (co-authored with B. Zelizer, 2010). His edited collections include *Citizen Journalism: Global Perspectives* (co-edited with E. Thorsen, 2009), *Rethinking Communication* (2010) and *The Routledge Companion to News and Journalism* (2010). Much of his current work revolves around crisis reporting, with a particular interest in citizen contributions – especially imagery – to news coverage.

SANJEEV CHATTERJEE

Sanjeev Chatterjee is Vice Dean, Associate Professor and Executive Director of the Knight Center for International Media at the University of Miami. Chatterjee is an award winning documentary filmmaker. He has taught classes in studio and field production, media and society, writing and documentary production at the University of Miami. He received an Excellence in Teaching Award in 2002 and has been nominated two more times since. Professor Chatterjee is producer, co-director and writer of a global motion picture project about potable water entitled "One Water" (http://www.onewater.org). An earlier short version of the film won two awards at the Broadcast Education Association and has been screened at special United Nations conferences in 2004 and 2005 as well as a special jury award at the World Water Forum in Mexico City in 2006. A feature version of "One Water" premiered on the closing night of the Miami International Film Festival in 2008.

MANUEL GUERRERO

Manuel Alejandro Guerrero is Dean of the Department of Communication and Director of Ibero 90.9 FM Radio at the Universidad Iberoamericana in Mexico City, member of the National System of Researchers, and Academic Coordinator of the Professional Electoral Service at the Federal Electoral Institute in Mexico. He holds a Ph.D. in Political and Social Science from the European University Institute in Florence, Italy and an M.Phil. in Latin American Studies from the University of Cambridge. His research has been focused on the role of the media in new democracies, especially on the framing of political issues, and on media and political attitudes and electoral behavior. On these topics he has a number of book chapters, journal articles, and two books.

GEORGE LUGALAMBI

George W. Lugalambi, PhD, is a media and communication analyst and researcher. A former journalist and newspaper editor, he was until May 2011 a senior lecturer and chair of the Department of Journalism and Communication at Makerere University, Uganda, where he did the work on this chapter. He has collaborated on the media literacy programs of the Salzburg Academy on Media and Global Change since 2008.

JAD MELKI

Jad Melki, Ph.D., is Assistant Professor of Journalism and Media Studies at the Media Studies Program at the American University of Beirut, Lebanon. He is also the research director of the International Center for Media and the Public Agenda and a faculty member at the Salzburg Academy on Media and Global Change. Previously, Melki was a visiting faculty at the graduate Communication program at Johns Hopkins University and at Towson University. He was also a project manager at the Salzburg Global Seminar and a faculty researcher at the Phillip Merrill College of Journalism at the University of Maryland, College Park, where he received his Ph.D. Melki teaches research methods, media and society, media literacy, broadcast and digital journalism, and media, war and terrorism. His research interests rotate around news media coverage of war and global issues, digital and social media, media habits of youth, Arab media and politics, Arab media education, and media literacy. Melki had been a broadcast and online journalist for over 10 years working with US and Arab media. He was part of the Webby award and Press Club award winning Hot Zone team (Yahoo! News) covering the 2006 Israeli war on Lebanon.

PAUL MIHAILIDIS

Paul Mihailidis is Assistant Professor in the Department of Marketing Communication at Emerson College in Boston, MA, and Director of the Salzburg Academy on Media and Global Change (www.salzburg.umd.edu). Mihailidis's research concerns the connections between media, education, and citizenship in the 21st Century. He has published widely on media literacy, global media, and digital citizenship. He is the editor of *News Literacy: Global Perspectives for the Newsroom and Classroom* (Peter Lang), co-author of the forthcoming *Media Literacy Learning Commons* (Libraries Unlimited), and co-editor of *Perspectives in Media Literacy*. Mihailidis sits on the board of directors for the *National Association for Media Literacy Education* (NAMLE), and is reviews editor for the *Journal of Media Literacy Education* (JMLE). At Emerson, Mihailidis teaches Interactive Communication, Understanding Consumers, Social Media, and Media Literacy.

SUSAN MOELLER

Susan D. Moeller is the director of the International Center for Media and the Public Agenda (ICMPA) at the University of Maryland, College Park. Moeller is also Professor of Media and International Affairs at the Philip Merrill College of Journalism and an affiliated faculty member of the School of Public Policy at Maryland. She is co-founder and a faculty member of the Salzburg Academy on Media & Global Change in Austria, an initiative of ICMPA and the Salzburg Global Seminar. Moeller is the author of a number of books, including *Packaging Terrorism: Co-opting the News for Politics and Profit* (2009), *Compassion Fatigue: How the Media Sell Disease, Famine, War and Death* (1999), and *Shooting War: Photography and the American Experience of Combat* (1989). She is the lead author and editor of a Freedom of Expression Toolkit and Model Curriculum for UNESCO's Division for Freedom of Expression, Democracy and Peace. Her commentary appears frequently in newspapers and magazines around the world and she blogs for the *Huffington Post*, the World Bank and *Foreign Policy* magazine.

CONSTANZA MUJICA

María Constanza Mujica Holley is a Chilean journalist from Pontificia Universidad Católica de Chile and PhD in Hispanoamerican Literature by that same university. Since 2001, she is a professor in the Faculty of Communications in Pontificia Universidad Católica de Chile. She has centered her research interests around two areas: discourse analysis of television programs, with special emphasis on the study of melodrama as narrative structure in journalism and telenovelas, and the study of journalistic quality. She has co-authored books and authored papers on melodrama, journalism quality and theory, and foreign news on television. Professor Mujica participated as Faculty in the Salzburg Academy on Media and Global Change on 2008 and 2009.

STEPHEN REESE

Stephen D. Reese is currently Jesse H. Jones Professor of Journalism and Associate Dean for Academic Affairs in the College of Communication at the University of Texas. At Texas he has taught a wide range of subjects, from broadcast news to critical thinking for journalists, and served as director of the School of Journalism. His research has focused on a wide range of issues concerning media effects and press performance--and been published in numerous book chapters and articles in, among others, *Journalism & Mass Communication Quarterly, Communication Research, Journal of Communication, Journal of Broadcasting & Electronic Media, Public Opinion Quarterly, Harvard Journal of Press-Politics, Journalism Studies,* and *Critical Studies in Mass Communication.* He is co-author, along with Pamela Shoemaker, of *Mediating the Message: Theories of Influence on Mass Media Content* and editor of *Framing Public Life: Perspectives on Media and Our Understanding of the Social World.* Reese has lectured internationally at universities in China, Colombia, Mexico, Spain, Germany and Finland, and was the Kurt Baschwitz Professor at the University of Amsterdam in 2004. He has been a regular summer faculty member at the Salzburg Academy on Media and Global Change.

STEPHEN SALYER

Stephen Salyer has been the president and chief executive officer since 2005 of the Salzburg Global Seminar, a policy and leadership development center in Salzburg, Austria. He credits attending a 1974 Salzburg Seminar on "The Social Impact of Mass Communications" as influencing his choice of a career in media and interest in how different societies frame and understand public issues. Salyer was CEO of Public Radio International, 1988-2005, and helped create and launch such public radio programs as Marketplace, The World, Studio 360, and others. From 1979-1988, he was a senior executive at WNET/Thirteen, the flagship public television station in New York City, in charge of program development, education, marketing and communications. In 1999, he co-founded Public Interactive, Inc., providing on-line services to hundreds of public radio and television stations in the United States, and served as its chairman until 2005. Salyer co-founded in 2007 the Salzburg Academy on Media and Global Change, an comparative program involving the world's leading journalism students and faculty in co-creating curricula and research on digital and social media. He serves as a director of Davidson College and Guidestar, Inc. He and his wife, Susan Moeller, divide their time between Washington and Europe.

RAQUEL SAN MARTÍN

Raquel San Martín is a professor at the Institute of Social Communication, Journalism, and Advertising at the Pontificia Universidad Católica Argentina in Buenos Aires. She teaches courses related to Journalistic Writing and Introduction to Journalism, and she is also member of the Advisory Board. She works as editor of Culture and Education in *La Nación*, a national newspaper in Argentina, where she also writes for the Literary Supplement. Her research interests include journalist's self-perceptions about their job, journalistic quality and the construction of journalistic values, such as objectivity. She is currently working on her research about the way in which crime and violence is depicted in the media, especially in newspapers. She is also interested in art history. She holds a degree in Journalism from Salvador University (Argentina), an MA in Journalism from Autonomous University of Barcelona (Spain) and is currently finishing an MA in Social and Political Anthropology from Latin American School of Social Sciences (FLACSO).

MOSES SHUMOW

Moses Shumow, Ph.D., is an assistant professor of journalism and broadcasting at Florida International University in Miami, FL. He teaches classes in global media, news literacy and multimedia production, and his research focuses on the intersections between immigration and media, with a particular focus on Latin America and the Caribbean. He also conducts research into the pedagogy of media production and information literacy and his work has been published in the International *Journal of Communication, Journal of Media Literacy Education* and *Taiwan Journal of Democracy*. Dr. Shumow graduated from the University of Miami, FL., in 2010; prior to that, he spent nearly a decade working in documentary filmmaking, helping to produce nationally televised programs for PBS, Discovery Channel/Discovery Español, National Geographic, History Channel, and Fox Sport en Español.

INDEX

Lee B. Becker, *General Editor*

The Mass Communication and Journalism series focuses on broad issues in mass communication, giving particular attention to those in
which journalism is prominent. Volumes in the series examine the
product of the full range of media organizations as well as individuals
engaged in various types of communication activities.

Each commissioned book deals in depth with a selected topic, raises
new issues about that topic, and provides a fuller understanding of it
through the new evidence provided. The series contains both single-
authored and edited works. For more information and submissions,
please contact:

Lee B. Becker, Series Editor | *lbbecker@uga.edu*
Mary Savigar, Acquisitions Editor | *mary.savigar@plang.com*

To order other books in this series, please contact our Customer Service
Department at:

> (800) 770-LANG (within the U.S.)
> (212) 647-7706 (outside the U.S.)
> (212) 647-7707 FAX

Or browse online by series at www.peterlang.com